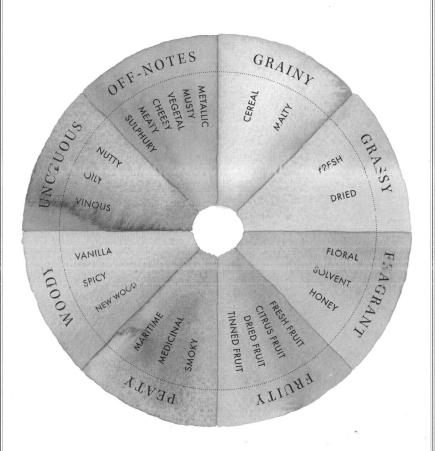

'Charles MacLean is Scotland's leading whisky expert.'
The Times

'Charles MacLean writes like no other expert on the subject. His prose is informed and highly entertaining.'
The Independent

'Charles MacLean is a world authority on malt whisky.'
Barbara Mowlem, *Daily Telegraph*

'Charles MacLean is, for my money, the most serious and scholarly writer on whisky today,'
Andrew Jefford, *London Evening Standard*

'He is the doyen of whisky writers . . . The fount of all knowledge on the subject.'
Tom Morton, BBC Scotland

'. . . acknowledged in his industry as both a great lover of and expert authority on Scotch whisky.'
Al Horsford, *Essential Medical World*

'MacLean, a moderately tall man of indeterminate age and stocky build, has an old world air about him – if one were to get precise with such allegories – of a former travelling minstrel, a sometime Smee-like pirate, now retired to the less precarious life of landed gentry. He has an impressive moustache of the Mark Twain/Rudyard Kipling school of whisker-growth, and on his head a mop of hair with a distinguished recession at the temples. At the time we met, he wore a monocle that dangled from a string around his neck (apparently his wife hates it) and a pair of pink pants that did nothing to take away from his overall genial gruffness.'
Vinayak Varma, *New Indian Express*

Charles MacLean's

Whiskypedia

A Gazetteer of Scotch Whisky

BIRLINN

First published in Great Britain in 2009 by Birlinn Ltd.
A revised edition published in 2014 by Birlinn Ltd.
This revised edition published in 2016 by Birlinn Ltd.

Birlinn Ltd
West Newington House, 10 Newington Road,
Edinburgh EH9 1QS

www.birlinn.co.uk

ISBN 978 1 78027 401 0

British Library Cataloguing-in-Publication Data
A catalogue record for this book is available on
request from the British Library.

Design and typesetting by 3btype.com
Cover design by Glen Tutssel

Printed and Bound in China

For Sheila MacLean

'At first I found my descriptive task beset
with difficulties far exceeding what I had
contemplated; but . . . my interest grew with
the work, making it easier for me; until,
after a time, I acquired quite a zest for
these distillery studies.'

ALFRED BARNARD, *THE WHISKY
DISTILLERIES OF THE UNITED KINGDOM,*
1887

Contents

Foreword

This is a book about Scotch whisky distilleries, but it is also a book about why Scotch whisky tastes as it does, and how flavour might have subtly changed over the years.

It embraces both malt and grain distilleries, including those which have closed since 1945, but whose product is still available, and those which have opened recently and whose product is not yet available. Each entry provides a brief account of the distillery's history, highlighting anything that I find especially interesting. A secondary text also supplies details of how the whisky is made there (today, and in the past), for those who are interested in such minutiae! With the help of the introductory 'Notes', I hope this will provide enthusiasts with some idea of what the make from each distillery tastes like – and why, by explaining how raw materials, plant, process and wood contribute to flavour.

My hope is that the book will provide a succinct and reliable account of each distillery and its whisky. I am aware that this has been done before – not least by myself! – but it is surprising how quickly things change in the whisky industry, and how soon whisky books become dated.

This is truer today than ever: despite the global recession, there has been a boom in demand for Scotch whisky, and this has led to a dramatic increase in distilling capacity. Since the first edition of *Whiskypedia* appeared in 2009, twelve new malt distilleries have opened; a staggering 39 more are currently proposed; 21 distilleries have changed ownership; capacity in many established distilleries has been increased – substantially in many cases (see p.13 for details).

Will demand keep pace with supply? Only time will tell. A friend with long experience in the whisky industry recently asked me: 'How do you make a small fortune out of a distillery?' The answer is: 'Start with a large one'!

I began writing about Scotch whisky in 1981 – a brochure for Bell's. Only in retrospect have I learned how the industry has changed since then. During the 1980s I was fortunate to be asked by several whisky companies to research and write materials for them – summary company histories, such as those for Highland Distilleries Centenary and Glenmorangie's Sesqui-Centenary; brand histories, such as those of the nascent United Distillers in 1988. By training and temperament I am an historian (during the same period I wrote five books on Scottish-historical subjects) and working for Scotch whisky companies brought together my passions for Scotland, its social history and our national drink.

By 1988 I had written enough about whisky to make a credible approach to a publisher for a book on the subject. This appeared five years later as *The Pocket Guide to Scotch*, in Mitchell Beazley's series of the same name. Following its success, a further three books were commissioned by other publishers, then Mitchell Beazley asked me to write a larger illustrated book. *Malt Whisky* was published in 1997 and won a Glenfiddich Award for its photography and design; it has since been reprinted five times and translated into eight languages.

My editor moved to Cassell and commissioned *Scotch Whisky: A Liquid History*. This took four years to research and write, but won 'Best Wine & Spirits Book' in the James Beard Awards (the first time in the history of these prestigious American awards that a spirits book has won!) and 'Best Drinks Book' in the World Food Media Awards. It was also runner up in the U.K.'s André Simon Awards.

A further eight titles have appeared since then – including *Maclean's Miscellany of Whisky* (now in its fourth edition), *World Whisky* (translated into ten languages), *Famous for a Reason* (a history of The Famous Grouse – the heaviest whisky book ever published!) and *Spirit of Place* (an illustrated account of leading malt whisky distilleries). Two more books are currently in progress.

As will be apparent, it takes a long time to write a book! One of the reasons is that one has to make money from other sources than book royalties. In my case this came from whisky companies which employed me to write or talk about my pet subject, to present tastings and host events around the world. I am eternally grateful to them. Without such support I would not have been able to continue my researches.

The first edition of *Whiskypedia* took four years to complete and since 2009 the book has been revised and updated three times. I hope it will be a cheerful companion for those of you who are starting out on the journey of discovery into the world of Scotch whisky, as much as for those who are already familiar with the subject, experienced in the nuances of flavour which distinguish the product of one distillery from another. I also hope the book will be useful to the ever increasing number of enthusiasts who collect whisky: where possible I have recorded when dramatic changes have been made to individual distilleries and when their owners began to release their products as single malts.

Notwithstanding this, at the end of the day, the purpose of Scotch whisky is to give pleasure, and such pleasure can be as simple or as subtle as you like – hence the Whisky Wheel inside the front cover and the expanded chapter on Appreciating Whisky. The glory of the drink is that it can be enjoyed at so many levels, and rewards consideration as well as merely being an intoxicant. It is, after all, the most complex spirit in the world.

Charles MacLean
Edinburgh

Historical Overview

Scotch whisky is currently experiencing a boom greater than at any time in the industry's history. Not since the 1890s has so much been invested in expanding production. Twenty new distilleries opened between 2004 and 2016*, and a further thirty-eight are proposed or under construction. Many well-known distilleries have been expanded, several doubling capacity, and two – The Glenlivet and Macallan – are building massive new distilleries on site, capable of producing 32 million and 17 million litres of pure alcohol per annum respectively. Over the past ten years malt whisky production capacity has increased by over 40%, from 2.46 million L.P.A. in 2006 to 3.47 million L.P.A. in 2016.

Such optimism is based upon the anticipated global demand for Scotch whisky over the coming decades – a remarkably difficult exercise, vulnerable to factors beyond the industry's control, including the global economy and international politics, not to mention sale-of-alcohol regulations, fiscal arrangements and fashion in over two hundred markets.

Let's hope the marketers have their projections right. The history of Scotch whisky is a story of booms and busts. The expansion of the 1780s terminated suddenly in 1788. Of the dozens of new distilleries that opened after 1823, only a small percentage survived 'the hungry '40s'. The great Whisky Boom of the 1890s turned dramatically to bust in 1900. Notwithstanding the global economic down-turn since 2008, Scotch has continued to prosper in export markets and is holding its own in the U.K., which has been a flat market for over a decade. The growth is led by blended

---

* Glengyle (2004), Kilchoman (2005), Glenburgie II (2005), Daftmill (2005), Ailsa Bay (2007), Roseisle (2008), Abhainn Dearg (2008), Starlaw (2010), Wolfburn (2013), Strathearn (2013), Annandale (2014), Arbikie (2014), Ardnamurchan (2014), Ballindalloch (2014), Kingsbarns (2014), Eden Mill (2014), Dalmunach (2015), Isle of Harris (2015), Glasgow (2015), Inchdairnie (2016)

Scotch, which still commands over 90% of the total market by volume, but where blended brands lead, malt whiskies follow, as the statistics demonstrate. (See *Facts and Figures* for the most recent statistics available.)

THE ORIGINAL SCOTCH

Whisky made in Scotland from malted barley is the original Scotch, although by the late eighteenth century, with the arrival of large-scale commercial distilling in the Lowlands, mixed grains (wheat and rye, as well as barley) were being used. With the invention of the continuous still in the late 1820s (perfected and patented by Aeneas Coffey, former Inspector of Excise in Dublin, in 1830), Lowland distillers gradually came to devote themselves to grain whisky production in such stills, producing very pure, very high-strength, somewhat bland spirit, with which they inundated the new industrial towns in the Central Belt of Scotland, as well as sending quantities to England for rectification into gin.

Pot still malt whisky was extremely variable in flavour. Like Longfellow's little girl, 'when she was good, she was very, very good and when she was bad she was horrid'! It was drunk straight and young in the Highlands, and mixed into a punch in the Lowlands, with water, sugar, lemons and sometimes spices. The first reference to the benefits of maturation that I know of is in Elizabeth Grant of Rothiemurchus's *Memoirs of a Highland Lady*, where she recalls sending 'pure Glenlivet whisky . . . long in the wood, long in uncorked bottles [this is mysterious!], mild as milk and with the real contraband *goût* [i.e. taste] in it' for the delectation of King George IV in Edinburgh in August 1822.

THE ECLIPSE OF MALT WHISKY

It is safe to suppose that wine and spirits merchants would have mixed the light grain whisky with variable and pungent malts to create a drink with broader appeal, from at least the 1820s. Many were familiar with blending teas, wines and cordials, which were often also part of their stock in trade. But the first branded blend appeared in 1853, Usher's Old Vatted Glenlivet, and after Gladstone's Spirits Act of 1860, which allowed the mixing of malt and grain whiskies under bond (i.e. before duty had to be paid), blended Scotch took off.

And take off it did – in a very big way – helped by a variety of factors, not least of which was the devastation of European vineyards by the louse *Phylloxera vastatrix*. By the late 1860s non-vintage cognac was unobtainable, and since 'brandy and soda' was the drink of the English middle classes, this caused considerable dismay. Blended Scotch (and soda) was now so improved as to be in a position to replace it.

Historians refer to the 1890s as the era of 'The Whisky Boom'. Henceforward the fortunes of malt whisky distillers would be inextricably linked to those of the blenders, and the leading blending houses built or bought or took an interest in malt distilleries in order to secure the fillings for their blends. Thirty three new distilleries were commissioned during the 1890s, 21 of them on Speyside (almost all of which are still in operation).

Alfred Barnard, the indefatigable Victorian traveller and editor of *Harper's* magazine, visited 118 malt distilleries during the mid 1880s while compiling his monumental book *The Whisky Distilleries of The United Kingdom* (1887) – the first and still the most thorough account of distilling in these islands.

With considerable foresight, he remarked: 'We shall be treading on delicate grounds when we refer to the fact that there are those who hold that the future of the Whisky trade lies with Malt Whisky. Certainly "the present" is not entirely in the hands of that product. Blenders without number can be found who will strenuously affirm that to give the public a moderate priced article with sufficient age, there is no way but to use good old Patent Still Grain Whisky as a basis.'

'Delicate grounds' indeed. Matters came to a head with the 'What is Whisky?' Inquiry, prompted by an English court in 1905 ruling that, to be named 'whisky', the spirit must be made in a pot still. This was followed by a Royal Commission

(1908–09), which found that 'The market for blended whiskies is greater than that for individual whiskies; so much so that it would probably be safe to say that the majority of Englishmen who drink whisky seldom drink anything but a blend', and, therefore, that patent still spirit made from grains other than malted barley, and a mix of this with straight malt [i.e. blended whisky] had equal rights to be called 'whisky'.

In truth, very little malt whisky was released as 'single' – Barnard remarks: 'there are only a few of the Scotch Distillers that turn out spirit for use as single whiskies' – and those single malts that did appear were usually bottled for local sale by hotels or spirits merchants, or bought in bulk by the small cask or stoneware jar by private customers.

In 1930 it was possible for Aeneas Macdonald to lament: 'the notion that we can possibly develop a palate for whisky is guaranteed to produce a smile of derision in any company except that of a few Scottish lairds, farmers, gamekeepers, and bailies, relics of a vanished age of gold when the vintages of the north had their students and lovers'. (*Whisky*, 1930)

BOOM AND BUST

Not long after he wrote this, the owner of The Glenlivet Distillery, Bill Smith Grant, began selling his make as a single malt on Pullman coaches in the United States, but this was the exception to the rule: 99.9% of the malt whisky made went into blends.

As will be apparent from a cursory reading of the individual distillery entries below, there was a huge growth in production in the two decades following the end of World War II. Scotch was again enjoying a boom period, embraced as 'the drink of the free world' in Europe as well as the U.S.A. and the U.K. Exports rose from 23.15 million proof gallons in 1960 to 107.08 million proof gallons in 1980; in the U.K. consumption tripled from 7.2 to 21.22 million proof gallons.

Three new grain distilleries and 26 new malt distilleries

were built between 1962 and 1972, four of them within existing grain distilleries and 11 alongside existing malt distilleries. The latter were built by Scottish Malt Distillers (S.M.D.), the malt distilling division of the Distillers Company Limited (D.C.L.), the largest whisky company, all to a similar design, the so-called 'Waterloo Street' style, named after S.M.D.'s headquarters in Glasgow (see *Caol Ila*). Many other distilleries were modernised and expanded, most to double capacity.

By the mid 1970s, economic conditions were less stable and the huge U.S. market had begun to contract. Moreover, in both the U.S.A. and the U.K., there was a shift in consumer taste towards blander spirits (white rum and vodka) drunk with mixers, and to wine; younger drinkers had been especially targeted and lifestyle advertising was designed to appeal to them. Scotch lost its fashionable cachet, and was now seen as 'Dad's drink'.

Whisky companies responded by exploring other markets, particularly in South America, Japan, Hong Kong and Europe, but production far outstripped demand. By 1980 the amount of whisky being held in bond was more than four times that in 1960. There was talk of a 'whisky loch', comparable to the 'wine lake' in France and the 'butter mountain' in northern Europe.

Furthermore, during 1981 the world economy slipped rapidly into recession. Output of both grain and malt whisky declined sharply (in 1983, it was at its lowest level since 1959). Between 1981 and 1986, no fewer than 29 distilleries were taken out of production; 18 of these have not worked since, and several have been demolished.

THE RENAISSANCE OF MALT WHISKY

Throughout all this period, single malt whisky remained relatively uncommon. Glen Grant and Glenfiddich were the exceptions.

Glen Grant was one of the few single malts widely available in the U.K., but it was in Italy during the 1960s that sales really took off, and this was owing to the efforts of the brand's distributor, Armando Giovinetti. Convinced that malt whisky would appeal to the Italian palate, he approached several distillers during the late 1950s, but was turned down: 'all our make goes for blending'. When he approached Glen Grant, Douglas McKessack, the owner and great-grandson of the founder, referred him to Charles H. Julian, the company's bottler in London. In 1960 he obtained 50 cases of 12YO; as he told me himself, 'My attitude was, if I don't sell it, I'll drink it.'

Within a year, and after visiting the distillery, he had decided that a younger whisky was required and obtained 100 cases from Julian. Through his efforts Italy became the leading export market for Glen Grant by 1970, and the first country in the world to embrace single malt Scotch whisky. By 1977 sales of Glen Grant were around 200,000 cases a year.

In 1964 the directors of William Grant & Sons, all descendants of the founder, decided to bottle their Glenfiddich Pure Malt as an eight-year-old, keep it pale in colour – like the successful Glen Grant – and present it in a dark green version of the characteristic triangular bottle used for their Standfast blend. The company also set about promoting the brand heavily, often with events and clever stunts that attracted editorial coverage and free publicity – the very novelty of 'pure malt' (it was labelled 'Straight Malt' in the U.S.A.) attracted interest. The company even struck on the simple but effective idea of supplying London's theatres and film studios with bottles of Glenfiddich filled with flat ginger ale as props, which tasted a whole lot better than cold tea! By 1970, the brand was selling 24,000 cases in the home market, and was being heavily marketed through the newly enfranchised airport duty-free shops.

William Grant & Sons effectively achieved a ten-year 'first mover advantage', to use a marketing term, over the other malt distillers, some of whom looked on with interest at their achievement in promoting their product as single malt, notably the independent companies Macallan-Glenlivet and Macdonald & Muir, owners of Glenmorangie distillery.

As early as 1963, the Chairman of Macallan, George Harbinson, reported that 'the sale of Macallan in bottle is gaining momentum with a steadily increasing demand for the over 15-year-old from the south of England', and again in 1965: 'the interest in single malts is undoubtedly increasing and larger sales are expected'. Next year, Messrs Fratelli Rinaldi of Bologna were appointed sole agents for The Macallan in Italy, and, following an advertising campaign, in 1967 they 'ordered more whisky than the total amount which went towards the home market'. Agents in France were appointed three years later.

The 1972 Annual Report noted that 'sales of cased Macallan had doubled during the year', and added, prophetically, that 'a large increase in this type of business was anticipated in light of a fantastic growth in public interest, which would eventually see malt whisky becoming extremely fashionable'. The directors decided to conserve stocks of mature whisky, even at the expense of demands from blenders, and to 'put larger proportions aside for selling in cases'.

In 1978 Macallan appointed its first marketing director, Hugh Mitcalf, who moved across from Glen Grant, following Seagram's takeover of that company, and who had been instrumental in the tremendous success of the brand in Italy. The year he arrived, the entire promotional budget allocated to The Macallan amounted to £50, but this was about to change. Witty advertising, tasteful repackaging and word of mouth soon made Macallan a household name.

Glenmorangie's story was similar, although unlike Macallan, which relied on outside companies (mostly Robertson & Baxter, its fillings agent) to buy its make, its owners, Macdonald & Muir of Leith, had a raft of blends of their own to supply, notably the popular Highland Queen range. Demand was heavy during the 1970s, and the distillery's capacity was doubled (to four stills) in 1977, but only a small amount was bottled as a single, and it was not supported by advertising until 1981.

That year the company allocated just under £200,000 to a print campaign in broadsheet newspapers. Compared with other whisky advertisements, the style and approach was novel, stressing the 'craft' elements that went into the making of the whisky and focusing on the fact that the distillery has the tallest stills in the industry. The strapline was 'A little nearer heaven than other malt whiskies'. The craft theme would be developed and extended in the 1980s and 1990s, with the long-running but hugely successful 'Sixteen Men of Tain' campaign.

ON THE BANDWAGON

Other companies began to follow suit, either bottling their malts for the first time or repackaging and using limited advertising. Justerini & Brooks launched Knockando in 1978 as a 12-year-old, with stated vintages. Highland Park, Tamdhu and Bunnahabhain were repackaged by Highland Distillers and relaunched in late 1979, and Highland Park rapidly took off. Possibly the most elegant repackaging of all was that for The Balvenie, by William Grant & Sons in 1982. Now named Founder's Reserve, without an age statement, it was filled into a long-necked vintage cognac bottle, unlike any other Scotch. Aberlour began to be heavily promoted in France around 1980 and a lavish booklet was produced in 1983 by Long John International to promote Tormore in the U.S.A. Glen Garioch was repackaged by Stanley P. Morrison around the same time, as was Auchentoshan. Bell's made Blair Athol, Inchgower, Bladnoch and Dufftown available for the first time. Whyte & Mackay promoted Jura and Tamnavulin. And so on.

Cardhu, owned by the mighty Distillers Company, had been available as a single since the 1960s. Now D.C.L. began to promote it with print advertising, stressing its rarity ('There are approximately 10,000 cases of Cardhu offered for sale each year. This is not a lot, and the whisky is, therefore, quite scarce.'); they even took the novel step of allowing press and V.I.P. visitors to the distillery. In 1982 Distillers launched 'The Ascot Malt Cellar' – Ascot was the company's home trade base – a collection of four singles and two vatted malts. By all accounts this was done reluctantly and was not widely promoted. It made no impact on the marketplace, but the seeds of exploring different styles and regional differences were already there, and this would be widely exploited in years to come.

Significantly, in 1988, D.C.L.'s successor, United Distillers, launched a range of malts that stressed 'regional differences' and opened up the whole sector. From their huge inventory of available whiskies, they selected six and named them 'The Classic Malts'. They were Lagavulin, Talisker (to represent the heavily peated style of Islay, and the slightly less smoky style of the Isles), Oban (maritime in character, representing the West Highlands), Dalwhinnie (a typical Highland malt), Cragganmore (a complex Speyside) and Glenkinchie (standing for the lighter Lowland style).

It was a move whose time had come, and was hugely successful. Other distillers attempted to follow suit. Allied introduced its 'Caledonian Malts' in 1991, featuring Tormore (a Speyside), Miltonduff (also a Speyside), Glendronach (from Aberdeenshire) and Laphroaig (from Islay), but it was not a success. Later Scapa (from Orkney) replaced Tormore, but the exercise was abandoned in 1994. That year Seagram tried the same idea with its 'Heritage Selection' – Longmorn, Glen Keith, Strathisla and Benriach – all good malts, but all from Speyside, and having a broadly similar flavour profile. The project did not last.

Some distillers who did not have a portfolio of regional malts to offer sought to extend their product range by offering their malts at different ages, or matured in different woods. Pioneers of this were Teacher's, who offered two 12-year-old versions of Glendronach in the late 1980s, one matured in sherry-wood, the other, described as the 'Original', in a mix of sherry and bourbon casks. About the same time,

Macdonald & Muir produced an 18-year-old expression of Glenmorangie, which also included some sherry casks in its mix.

In 1994 the latter went a step further, launching an expression of Glenmorangie which had been reracked into port casks for the final months of its maturation, a process known as 'finishing'. This was followed by Madeira- and sherry-finishes (1996), and subsequently by many other expressions, mostly in limited editions. At exactly the same time as the Glenmorangie Port-wood appeared, William Grant & Sons issued an expression of Balvenie 12-year-old finished in Oloroso casks and named 'Doublewood'. Since the mid 1990s several brand owners have followed Glenmorangie's and Balvenie's lead.

Further variations could be introduced by bottling only a single cask, usually straight from the wood, at 'cask strength' and without chill-filtration. The latter is a technique to remove compounds that precipitate when water or ice is added, causing the whisky to go slightly hazy. Chill-filtration 'polishes' the spirit and prevents haze, but the compounds it removes make a big contribution to flavour and texture, so connoisseurs prefer them left in!

Single cask bottlings are often done by independent bottlers, the number of which increased dramatically in the 1990s. (See 'Leading Independent Bottlers' in *Facts and Figures*.)

Never in history have so many expressions of malt whisky been available to the consumer. Never in history have there been so many enthusiasts for the drink! Demand has led to shortages of certain makes in certain markets, and to various ways of solving the problem. At its simplest, the price goes up. Or, the brand is pulled from specific markets to supply others. This happened to The Macallan in Taiwan in the early 21st century, where demand for the 'traditional' style of sherried malt was met by moving stocks from other markets and replacing them with a parallel range of ages, matured in ex-bourbon casks, of which the company had plenty.

Famously, Diageo attempted to meet the demand for Cardhu in Spain by introducing 'Cardhu Pure Malt', a mix of Cardhu and Glendullan. The industry was outraged – or at least Diageo's competitors were, principally William Grant & Sons, who raised a storm of protest, maintaining that consumers were being conned, and that the reputation of single malt Scotch would be tarnished. 'Pure Malt' was withdrawn, and the Scotch Whisky Association set up a committee to look into definitions (see *Understanding the Label*).

Little is left of the surplus stocks of whisky built up during the late 1970s and mid 1980s – the so-called 'whisky loch' mentioned above – and malts of 30 and 40 years old now fetch eye-watering prices. There is also a shortage of younger malt, between 12 and 18 years old, owing to limited production between 1998 and 2004, with a view to balancing stock. Many whisky companies have responded to the pressure on stocks by removing age statements, arguing that it is better for consumers to make up their own minds, without resorting to age as a guarantee of quality.

Consumers are not happy about this. We know that advanced age is not an *absolute* guarantee of quality – we might even say that some malts are better drunk young – and we are increasingly aware of the fundamental role played by the cask in developing flavour: acceptable maturity can be reached in only few years in a highly active first-fill cask. But for decades, the pricing of older malts and aged blends has led us to believe that 'older is better'. In 2010 Chivas Bros even ran a global campaign for The Glenlivet under the headline 'Age Matters', but have now removed the age statements in the brand's core range.

In recent years Scotch whisky has been enjoying a boom period. Demand, particularly for old malts and deluxe blends, is greater than supply, so it is hardly surprising that prices have increased substantially in recent years. Too often, however, the new NAS (no age statement) expressions come with higher price tags yet do not reward comparison with their aged equivalents.

But the industry cannot be complacent. Since 2011 there has been a dip in the steady rise in volume and value shipments. However, the most recent figures (2015) allow room for optimism, particularly in relation to Scotch malt whisky, and there are many reasons to be confident about the future of Scotch. As I mentioned at the start of this essay, stock levels have increased massively, in anticipation of future demand. The scientific understanding of the key contributors to flavour makes for a higher quality of spirit. The arrival of so many new distilleries will expand our choice of malts. And the fact that Scotch has regained its reputation for style, flavour and versatility in so many global markets should lead to ever-growing demand.

Appreciating Whisky

The way you choose to drink whisky is the way you *enjoy* it most – straight, with water, ice, soda, lemonade, ginger ale, cola, green tea, coconut water (Dave Broom's recent book *Whisky: The Manual* (2014) provides a useful evaluation of different whiskies with some of these mixers).

Enjoyment is one thing; *appreciation* another. While the former is principally to do with taste and effect, appreciation engages all our senses – sight, smell, taste, touch (i.e. texture), even, some would say, hearing.[1] This is why, in the trade, the procedure is termed 'sensory evaluation' or 'organoleptic assessment'. But of the five senses, far and away the most important for assessing whisky is smell.

PROCEDURE

The standard procedure for assessing a whisky's character is simple:

1   Appearance

Principally this relates to colour (refer to the Flavour Wheel. The descriptors are my own; the numbers are those of the Lovibond scale and the European Brewing Convention (EBC) scale). You might also consider clarity, viscosity and viscimetry (see my *Miscellany of Whisky* for an account of the latter).

2   Aroma

The liquid is nosed straight initially, when you note the physical effect the alcohol vapour has on your nose. Then you add a little water – just enough to remove any prickle or burn, and to 'open up' the whisky, releasing aromatic volatiles. Whisky blenders work at 20%ABV; I prefer around 30%ABV – and nose again.

1   I know a distillery manager who can tell the difference between whisky which has been chill-filtered and that which has not by the sound it makes as it pours. There is also the undeniable sensory pleasure of cracking open or drawing the cork of a new bottle!

3   Taste

Remember that 'flavour' is, by definition, a combination of smell, taste and texture. It is not simply taste. What physical effect is the liquid having – smooth, oily, ascerbic, etc.? Now evaluate the balance of primary tastes across your tongue. Broadly speaking, sweetness is collected by receptors on the tip of the tongue: acidity and saltiness at the sides, bitterness at the back. 'Dryness' and 'smokiness' – not primary tastes – are detected as you swallow.

4   Finish

Consider the length of time you continue to taste or feel the spirit: is the finish long, medium or short? Note any lingering aftertaste.

THE SENSE OF SMELL

Compared with sight and taste, our sense of smell is infinitely more acute.

- While there are only three primary colours (blue, red and yellow, from which we construct our entire visual universe) and five primary tastes (sweet, sour, salty, bitter and umami – loosely defined as savoury), it has been estimated that there are 32 'primary' aromas.

- We are equipped with around 9,000 taste buds, but we have between 50 and 100 million olfactory receptors.

- We can detect odours in miniscule amounts: commonly in parts per million, in some compounds in parts per billion and in certain chemicals (some of them found in whisky) in parts per trillion. To grasp the enormity of this it is useful to think in terms of distance: one part per million is equal to one centimetre in ten kilometres; one part per billion to one centimetre in 10,000 kilometres (a quarter of the way around the world!); and one part per trillion a staggering one centimetre in ten million kilometres (or 250 times around the world!).

CONGENERS

The volatile, odour-bearing molecules found in whisky (and other liquids) are called congeners. Sensory scientists have identified over 300 in whisky, and they suspect there are as many again which have still to be isolated and described.[2]

It is these congeners which allow us to distinguish one whisky from another – and to distinguish whisky from brandy or vodka or wine – yet they make up only 0.3% of the contents of the liquid (the remainder being water and ethyl alcohol, both of which are odourless). This is equivalent to the depth of the meniscus in the neck of an unopened 70cl bottle. Vodka, a much purer – and therefore less aromatic – spirit than whisky, contains only 0.03% congeners, yet we can distinguish one vodka from another by its smell.

EMOTION

Smell is collected by receptors in the Olfactory Epithelium, a mucous-covered patch, above and behind the nose, which traps odour molecules and sends messages to the brain via the olfactory nerves. It is not known precisely how these receptors work, but the neural pathways from the Olfactory Epithelium connect directly to the lower brain, without being mediated by other receptor cells, as with taste buds.

The lower brain was the earliest part to evolve; it is referred to as 'Paleo-mammalian', i.e. it formed when we were still reptiles! – and includes the Limbic system, the seat of our long-term memory and emotions. This is why smell is the most evocative of our senses.

Scents can bring feelings and images flooding vividly back. The smells of childhood Christmases (the pine-needle smell of the tree, raw and cooked Christmas cake, mince pies, spices, mulled wine, candles), the smells of school-days (floor polish, disinfectant, chalk, sweat, carbolic

2   Coincidentally, it is reckoned that there are around 300 volatile odour-bearing compounds in human scent – a personal olfactory signature, picked up by sniffer dogs.

soap, sweets of all kinds), the smells of Guy Fawkes Night (fireworks, cordite, wood smoke), traditional household smells (cooking and baking, wax polish, cleaning products, coal or wood fires), smells associated with foreign holidays (markets and bazaars, foreign food, hot sand, tropical forests). Barbecues and parties; flowers and perfumes; tobacco, car interiors, seaside smells, country smells . . .

Pleasant and unpleasant smells – there are no pejoratives when describing smells! And even smells you have never smelled before. Among the many remarkable attributes of smell is that we can identify smells we have never encountered, although we may have difficulty in describing them.

DESCRIBING SMELLS

Putting words to smells is often difficult and takes practice, but it is hugely rewarding, since the very attempt to isolate and identify an odour focuses the mind, raises awareness and stimulates appreciation. It is also an essential part of the fun of enjoying whisky (or wine) with friends.

There is no fixed whisky vocabulary – even less so than wine, which connoisseurs have been trying to describe for far longer. The descriptors used in aroma wheels for whisky, such as the one in this book, while being generally similar and usually embracing the same key groups of aroma (the 'cardinal aromas' as the hub of the wheel) are for guidance only.

When nosing a whisky, I find it useful to run through the cardinal aromas, asking myself: 'Do I detect any cereal notes? Any fruity notes? Any peaty notes?' Et cetera. If I do detect, say, fruity scents, I move on to the next tier: 'Are they the scents of fresh, tinned, dried, cooked fruits?' Then I try to nail the precise fruit.

OBJECTIVE ANALYSIS

Objective analysis sets out to describe only 'what is there', limiting, as far as possible, the interpretative faculties of the person or panel doing the analysis. This is the kind of analysis done in the laboratories of whisky companies, and since the audience for such communications are colleagues, the vocabulary used is very limited and often derived from chemistry.

It is not necessarily descriptive: those with responsibility for assessing the quality and consistency of whisky need only to be certain that they are singing from the same hymn-sheet, not that their language means anything to outsiders. Thus, when they describe a sample as 'grassy' or 'meaty', they must be certain only that they all mean the same thing, even though these descriptors might not be the words that immediately spring to mind were outsiders to smell the same samples.

The vocabulary varies slightly from company to company. As an example, here is a list of the words used by Diageo's sensory experts to assess the character of new-make spirit:

Butyric	Nutty-spicy	Grassy
Peaty	Vegetal	Fruity
Sulphury	Waxy	Perfumed
Meaty	Green-oily	Clean
Metallic	Sweet	

SUBJECTIVE ANALYSIS

Subjective analysis gives free rein to the experience, imagination and recollection of the individual, limited only by the opinion of the rest of the panel.

The language is descriptive and figurative; it abounds in similes (which compare one aroma with another: 'smells like old socks', 'reminiscent of petrol') and metaphors (where an aroma is described in terms of what it resembles, not what it actually is: 'wood-smoke and lavender', 'distant bonfire on a seaweed-strewn beach').

It also makes use of abstract terms, such as smooth, clean, fresh, coarse, heavy, light, rich, mellow, young, etc. – and these, usefully, give rise to contrasting pairs: 'smooth/rough', 'clean/dirty', 'fresh/stale', etc. But often abstract terms are relative ('smooth compared with what?') and sometimes double meanings are possible ('young' = immature; 'young' = supple, lithe, well shaped). In general, abstract terms are useful for overall impressions – the general style and character of a whisky, its 'construction' and quality – rather than to describe specific aromas.

The language of subjective analysis can also be rhetorical – used to persuade or sell – but the effectiveness of the rhetoric depends upon the audience understanding the allusion. There is no point in describing a whisky as 'minty as a box of After Eights' to a customer who is unfamiliar with these chocolates!

CHEMICAL FOUNDATIONS

Smells have an objective existence as volatile aroma-bearing molecules. They are not figments of our imagination. We are all similarly equipped to identify them, although some of us are more sensitive to certain aromas than others, and some suffer from a degree of 'odour blindness', called 'specific anosmia' (total anosmia is rare).

Our noses are the ultimate arbiters of quality in whisky, and although there are expensive scientific instruments available to help quality-control departments with their objective analyses, no device has yet been invented which is as sensitive as the human nose.[3] Sensory chemists acknowledge the supremacy of the nose, and sometimes think of it as a 'human instrument' which can be trained and calibrated. But the calibrations are of our own making when it comes to subjective analysis.

The fact that scents are founded in chemistry provides a scientific basis for objective descriptors, but is also validates subjective descriptions and figurative language.

3   These instruments are called Liquid Gas Chromatographs. They register smells by measuring the amounts of volatile compounds present in a sample, and communicate the information graphically.

At its best, the language used to describe the aroma and taste of whisky is both colourful and objective; at worst it is marketing hyperbole, bearing little relationship to the product described. Increasingly, the tasting notes addressed to connoisseurs or customers buying from reputable specialist retailers are detailed, accurate and helpful – while also being entertaining and appetising.

Understanding the Label

Scotch whisky is the most tightly defined spirit in the world – for good reason: its popularity and high reputation make it an obvious target for counterfeiters. The Scotch Whisky Association (S.W.A.), the industry's governing body, is engaged in around 70 counterfeit actions around the world at any one time!

After two years of consultation, S.W.A. recommended to the U.K. Government and the European Commission that the definition be tightened, so as 'to ensure that consumers have clear information about what they are buying'. Its recommendations were embodied in The Scotch Whisky Regulations, in November 2009.

LIMITED EDITION
N o · **0000/2970**

Cₐ

ESTᴰ 1830

❶ **TALISK**

❷ **SINGLE MALT SCOTCH**

❺ **30 YEARS OLD**

❽ NATURAL CASK STRENGTH TALIS

❼

❼   ❾
49.5% vol 70cl ℮

THE O
WHISK

1 *Brand Name* is almost always the name of the distillery in which the whisky was made. The new rules prohibit the use of a distillery name other than the distillery in which it was made (e.g. a defunct distillery), with certain exceptions.

2 *Single Malt* is the product of an individual distillery. There is no such thing as a 'double malt', except in the glass!

3 *Scotch Whisky* must be made and matured in oak casks in Scotland for at least three years. New-make spirit is not 'Scotch'.

4 *Isle of Skye*, etc. indicates the region in which the distillery is located (see *Regional Differences*).

5 *Age Statement.* If the bottle has this, it is the age of the youngest whisky in the mix. It is permissible to add older whiskies.

6 *Date Bottled* or *Date Distilled.* Not all whiskies carry this, but it is useful if read against the age statement. However, whisky does not continue to mature in the bottle.

7 *Cask Strength.* The standard bottling strength is 40% or 43% Alcohol By Volume (A.B.V. or simply Vol.). Where the label says 'cask strength' it means it has been bottled without dilution, typically at around 60% A.B.V., although the term is not defined by law.

8 *Natural* or *Non chill-filtered.* Often associated with 'Cask Strength'. Most whisky undergoes chill-filtration on the bottling line to make sure it remains bright and clear when ice or water is added. This takes out certain compounds, which many connoisseurs prefer left in.

9 *Capacity.* 70cl is now the standard bottle-size in the E.U. Until 1990 it was 75cl, and it remains so in certain markets.

10 *Bottler.* This may be the distillery owner or an independent bottler. Under the new regulations, after 23.11.12 Scotch Malt Whisky must be bottled and labelled in Scotland. This was not formerly the case.

+ *Single Cask.* Bottled from an individual cask: a rarity – most single malts are vattings of many casks. (Not shown on this label.)

+ *'Wood Finished'* or *'Sherry/Port Finished'*, or *'Double Matured'* etc. tells you that the whisky has been reracked into different casks for the final months or years of its maturation. (Not shown on this label.)

10 TALISKER DISTILLERY
T, ISLE OF SKYE, IV47 8SR
4

3

6 bottled in **2008**
SINGLE MALT SCOTCH
OM THE ISLE OF SKYE
31030773

A Taxonomy of Scotch

(*with apologies to Linnaeus*)

KINGDOM:	Alcoholic Drinks
ORDER:	Distilled Beverages (as opposed to Fermented Beverages)
FAMILY:	WHISKY (Not vodka, brandy, rum etc.)
GENUS:	SCOTCH (Not Irish, American, Japanese etc.)
SPECIES:	SINGLE MALT SCOTCH WHISKY the product of an individual malt whisky distillery
	BLENDED MALT SCOTCH WHISKY a mix of malt whiskies from more than one distillery
	SINGLE GRAIN SCOTCH WHISKY the product of an individual grain whisky distillery
	BLENDED GRAIN SCOTCH WHISKY a mix of grain whiskies from more than one distillery
	BLENDED SCOTCH WHISKY a mix of malt and grain whiskies
GEOGRAPHICAL VARIETIES:	The Scotch Whisky Regulations 2009 require that only the five 'traditional regional names' – 'Highland', 'Lowland', 'Speyside', 'Campbeltown' and 'Islay' – may be used
MONO-SPECIFIC VARIETIES:	'Cask Strength' and 'Non chill-filtered'

The following abbreviations are used throughout this book:

D.C.L.	The Distillers Company Limited
I.D.V.	Independent Distillers and Vintners
S.M.D.	Scottish Malt Distillers
S.G.D.	Scottish Grain Distillers
U.D.	United Distillers
U.D.V.	United Distillers and Vinters

Regional Differences

Over the past 25 years it has become common to approach malt whiskies in terms of their regional differences, on the principle that the malts made in one part of Scotland are different to those made in another.

It was a brilliant marketing idea, which effectively communicated the fact that all malts are different in a way which was familiar to wine drinkers, used to the idea of differences within regions such as Bordeaux. The concept also made malts more accessible and encouraged consumers to explore whiskies from different parts of Scotland. What's more, it was justified by history in the ancient division between 'Highland' and 'Lowland' regions (first introduced for tax reasons in the 1780s), and the later identification of different styles from Islay, Campbeltown and 'Speyside' (originally called 'Glenlivet') by the end of the nineteenth century.

The first faltering step in the direction of marketing a selection of whiskies which displayed regional differences was made by the Distillers Company Limited when it launched – or, rather, slipped discreetly onto the market – 'The Ascot Malt Cellar' in 1982. This was a collection of six whiskies: Rosebank (a Lowland malt), Linkwood (from Speyside), Talisker (from the Isle of Skye), Lagavulin (from Islay), Strathconnan (a blended malt) and Glenleven (also a blended malt).

The idea was extended by D.C.L.'s successor, United Distillers, in 1988 when the 'Classic Malts' range was launched, specifically to demonstrate the differences between malts made in one region and another. All come from small, traditional, picturesque distilleries: the company had in mind the fact that consumers would want to visit the places where the malts were made, and soon developed visitor facilities at each.

It was an intelligent and appropriate move, and was hugely successful. Everyone began to talk about malts in terms of region – 'a cheeky little Lowland', 'rather a good example of the North Highland style', 'classic Perthshire whisky, of the sherried kind'. Writers divided and sub-divided the country into smaller and smaller 'regions'. Actually, in my own experience, this was driven by publishers, for whom a 'regional' or even 'sub-regional' breakdown of malt whiskies was more accessible and made for a more attractive book.

As we have seen, other whisky companies began to imitate the concept, but not having United Distillers' huge inventory of distilleries, their attempts were half-hearted.

THE LIMITS OF REGIONALITY

Although useful, the concept of regional differences is by no means infallible. If Islay malts are famously smoky – 'the most pungent whiskies made' – why are typically Bruichladdich and Bunnahabhain so mild, with not a trace of smoke? Lowlands are typically dry and short in the finish, yet Auchentoshan can be sweetly fruity and medium length. Speyside, where two-thirds of today's malt whisky distilleries are located, embraces a wide range of styles, light-, medium- and full-bodied. On top of this there's the influence of the cask during maturation, which can supply up to 80% of the flavour of mature whisky.

The truth is that it is not location – *terroir* – that dictates the style of a whisky, it is tradition. For 150 years, the main customers for malt whisky have been the blending houses, and the last thing they want when they buy malt x from Islay or malt y from Speyside is for those malts to have changed in style – if they have done, it messes up the formula for the blend.

The style, character and flavour of any malt whisky come from two principal sources: the way it is made and the way it is matured.

The way it is made embraces the peating levels in the malt, the way the malt is processed, how long the wash is fermented, the size of the stills and how they are operated, the style of the condenser (traditional worm tub or 'shell-and-tube') – above all, the craft of whisky making: how each individual distillery makes its whisky, and has been making it for decades.

The way it is matured encompasses the length of time the whisky has lain to the wood and the nature of the casks used – American or European oak? First-fill or refill? The kind of warehouse in which the whisky is matured (traditional or modern) and the location of the warehouse also play a part – but it's a small part, in relation to all the other factors.

Because of all these elements, I have included an account of plant and wood policy under each of the distillery entries, and an idea of the style of the new-make spirit and the mature whisky.

If they choose to, the owner of distillery *x* (on Islay) can make a 'Highland' style malt (and they do: see *Caol Ila, Ardbeg, Bruichladdich*), while distillery *y* (on Speyside) can produce a smoky 'Islay' style (and they do: see *BenRiach, Benromach*) – although neither could produce a replica of the make of another distillery, even if that distillery was next door.

But if the main customers for these distilleries are blenders – and remember that over 90% of the malt whisky made goes for blending – Mr X and Mr Y run the risk of losing their key customers. So they tend to stick to the way they have always done it, and to the extent that this way has been the way it has always been done on Islay or Speyside, or wherever, a local style emerges, a 'regional difference'.

Key to Entries

DISTILLERY

The name is that of the distillery, not the brand of single malt/grain whisky.

In most cases the two are the same; however, there are exceptions (Ballechin from Edradour Distillery, Longrow from Springbank Distillery, Old Pulteney from Pulteney Distillery etc.).

STATUS

All currently in operation, unless indicated:

dismantled buildings still stand, but plant gone

demolished buildings gone as well

conversion buildings are now used for something else

museum only one – Dallas Dhu

REGION

The regions adopted (per the current Scotch Whisky Regulation 2009) are: Lowland, Speyside, Highland (Island in brackets), Campbeltown, Islay.

ADDRESS, PHONE, WEBSITE, VISITORS

All self-explanatory.

OWNER

In some cases, the owning company or licensee is a subsidiary of a larger international concern (see 'Who Owns Whom?', in *Facts and Figures*).

CAPACITY

m L.P.A. = millions of litres of pure alcohol per annum

Many distilleries today are operating at capacity, but they vary production levels according to their stock levels. In 2002, for example, the industry was operating at merely 65.8% capacity; by 2008 this had risen to 91.1%, in 2013 94.4%, in 2015 85.4%.*

* Source: The Scotch Whisky Industry Review 2015

The spirit made in both malt and grain distilleries in Scotland cannot be called 'Scotch whisky' until it has matured for three years. Annual production – usually fixed at the beginning of the year – is geared to the requirements of customers, who are mainly the blending houses. Currently around 90% of the malt whisky made, and more or less 100% of grain whisky, goes into blends.

Only a small amount of the make of a malt whisky distillery (typically around 10% . . . although there are now many exceptions, where the whole output is bottled as single) is allocated for single malt bottling. So distilleries gear production to the needs of the blenders, their principal customers.

On their side, the blenders have to anticipate what the demand for their individual brands will be in 5, 8, 10, 12, 18, 25 (etc.) years' time. *In every market in the world, in which their brand has a presence.* A complicated forecast!

In truth, the availability today of single malts of advanced age may be attributed to the fall in demand for blended Scotch during the 1970s and 1980s.

HISTORICAL NOTES AND CURIOSITIES

Self-explanatory, and should be read in conjunction with 'Historical Overview' (see pp. 15–26).

WHERE DOES FLAVOUR COME FROM?

Nobody knows for *certain* – which is why Scotch is such a rewarding subject for study! And also why you can't make precisely the same flavour on a different site. Over the last 25 years, our knowledge of what influences flavour has increased hugely.

It was long thought that the location of the distillery (particularly a malt whisky distillery) was essential. 'Regional differences' between the styles of malt made in various parts of Scotland were discerned. Thus, many (but not all) Islay malts are smoky; Speyside malts tend to be sweet; Lowland malts are generally lighter in style, etc.

In fact you can make a smoky malt anywhere (see 'Malt' below) – and some Speyside distilleries are doing so; if you start saving spirit earlier, your make will be sweeter; if you run your stills and condensers in such a way as to encourage contact with copper, you will produce a lighter style of spirit (see 'Stills' below) – and so on.

Having said this, broad regional characteristics are discernible, not because distilleries can't change the style of their make, but because they don't want to, and neither do their customers, the blending houses. It is difficult to make a consistent blend if the component malt and grain whiskies are variable. (See 'Regional Differences'.)

Flavour comes from three things: raw materials, production and maturation. Each entry details aspects of each.

RAW MATERIALS

Water

Copious amounts of water are essential for distilling. It is used for production ('process water'), for cooling the condensers ('cooling water'), for reducing the strength of the spirit prior to filling into cask ('reduction water') and for the boilers that create the steam to heat the pans and coils in indirect-fired stills (see below).

In most cases one source of water supplies all these needs; in others, process and cooling water come from different sources.

It used to be thought that the nature of the water, its mineral content and acidity (i.e. hard or soft, and whether it was peaty or not) made a crucial contribution to the flavour of the whisky. The general view today is that this is not the case, although many distillers still believe it makes a small contribution, so I have indicated 'hard' or 'soft' in the notes.

The temperature of the water in the condensers certainly makes a difference; some distilleries choose to run them 'hot' or 'cold'; in the past, distillery operators could judge whether the spirit was made during summer or winter (the former was lighter in character than the latter).

Malt

Until the 1960s most distilleries malted their own barley on site, and where possible I have indicated when such maltings closed. Increased production at this time made it necessary for distillers to buy in malt from large centralised maltings, mostly owned by independent maltsters (Diageo has its own maltings), and this practice continues today.

Today, only eight distilleries make their own malt: Glen Ord has large maltings attached; Springbank makes all its requirement; Highland Park, Laphroaig, Bowmore, Kilchoman, Ardnamurchan, BenRiach and Balvenie have old-fashioned floor maltings to supply a proportion of their needs.

The variety of barley used to make the malt is not considered to affect flavour (although some would disagree!), but varieties change every few years, so I have not attempted to specify them.

If peat is burned during the drying stage, it will impart a smoky flavour to the whisky. Distillers specify the degree of peating they want, measured in 'parts per million phenols' (ppm). Phenols are the chemical compounds that impart smokiness or medicinal characteristics: lightly peated = around 2ppm, medium peated = around 15ppm, heavily peated = 25–35ppm, or in rare cases more than this.

PLANT

Mash tun

The 'mash' is a mix of ground malt (called 'grist') and hot water; the 'mash tun' is the large tub with a perforated base (made from cast iron, mild steel or stainless steel; usually covered, occasionally open) where this takes place, and where starches in the malt are converted into sugars by the action of enzymes.

The resulting liquid is called 'worts'; it filters through the bed formed by husks in the grist and may be 'clear' or 'cloudy' (i.e. containing small particles of starch). Cloudy worts give rise to a uniform 'maltiness' or 'nuttiness' in the spirit.

Until the 1970s, Scotch distillers used 'traditional' infusion mash tuns, called 'rake-and-plough' or 'rake-and-pinion' tuns. These have a rotating arm, equipped with revolving rakes to disturb the bed of the mash and facilitate draining.

During the expansion of the 1960s and 1970s, many distillers went over to 'Lauter tuns', invented by German brewers. In these, the rakes are replaced by small fins or 'knives'. In the case of 'semi-Lauter' tuns, the arms merely rotate; in 'full Lauter' tuns the fins may also be moved up or down, to take account of the depth of the bed.

The relevance of the style of mash tun is that it is difficult to achieve clear worts with a semi-Lauter tun, so I have indicated which kind is used in the individual distillery. I have also indicated its size, by noting the weight of grist used in each mashing.

Washbacks

A 'washback' is a fermentation vessel. Traditionally they were made from wood (principally Oregon pine/Douglas fir); many distillers now use stainless steel washbacks. Chemists deny that the material makes any contribution to flavour, although those who use wooden washbacks maintain that bacteria lurking in the wood assist secondary fermentations (they cannot be cleaned as rigorously as steel vessels), and also that they insulate the wash in cold weather.

Although I have not listed it, the length of the fermentation time makes a very important contribution to flavour: short fermentations (around 48 hours) make for a malty spirit; long fermentations (60–70 hours) develop fruity and floral flavours.

SPIRIT STILLS, DUNDASHILL DISTILLERY.

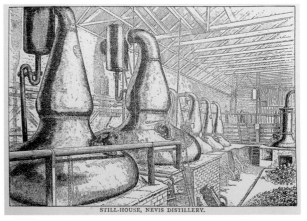

STILL-HOUSE, NEVIS DISTILLERY.

Still design has changed little since the nineteenth century, but stillhouses are tidier places since the stills ceased to be fired directly, by coal.

Stills

Malt whisky is always distilled twice – sometimes two and a half times, sometimes three times – so in most distilleries the stills are arranged in pairs, called the 'wash still' and the 'low wines' or 'spirit still'.

Stills come in three principal shapes: plain (or 'onion'), boil-pot (or 'boil-ball') and lamp-glass (or 'ogee'). There are many variations in size – indicated in the text by the amount of liquid charging each still – and in height. The still is connected to the condenser by the lyne arm (or lye pipe).

Stills must be made from copper; this metal acts as a purifier, removing unwanted sulphury compounds. The more contact the alcohol vapour has with copper (e.g. if the still is very tall, or if it is fitted with a purifier, or if it is run very slowly), the lighter and purer the spirit will be.

Some stills are fitted with 'after-coolers' (a second condenser) or 'purifiers' (a pipe from the lyne arm, which allows condensed vapour to return to the still for redistillation).

Condensers

The condensers continue the 'purification' theme. They are of two kinds: 'worm tubs' and 'shell-and-tube condensers'.

Worm tubs were the traditional condensers, and were (almost) universal until the 1960s. The 'worm' is a serpentine copper pipe, of gradually diminishing diameter. It sits in the 'tub' – an open vat of cold water. Copper contact in a worm tub is limited, so the resulting spirit tends to be heavier and richer than that coming from a condenser.

The 'shell' in a 'shell-and-tube condenser' is a copper or stainless steel column, within which is mounted a hundred or so small-bore copper pipes (the 'tubes'), through which flows cold water. The alcohol vapour enters the top of the column and condenses on the cold tubes, to be drawn off at the bottom. This kind of condenser increases copper contact, so makes for a lighter spirit.

Some stills are fitted with 'after-coolers' (a second

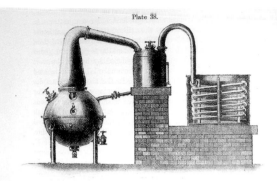

Plate 38.

COPPER STEAM-JACKETED STILL, WITH RETORT AND CONDENSING COILS. BLAIR,
CAMPBELL, & M'LEAN, LTD., WOODVILLE STREET, GOVAN, GLASGOW.

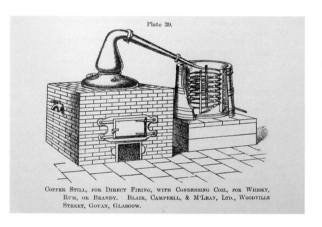

Plate 39.

COPPER STILL, FOR DIRECT FIRING, WITH CONDENSING COIL, FOR WHISKY,
RUM, OR BRANDY. BLAIR, CAMPBELL, & M'LEAN, LTD., WOODVILLE
STREET, GOVAN, GLASGOW.

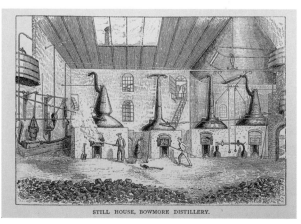

STILL HOUSE, BOWMORE DISTILLERY.

The top illustration is an 'onion' or 'plain' still shape, indirect fired by steam; the middle a 'lamp-glass' style; the bottom a mix of the two, with worm tubs beyond the back wall.

condenser) or 'purifiers' (a pipe from the lyne arm, which allows condensed vapour to return to the still for redistillation, producing a lighter style of spirit).

MATURATION

Casks

The cask in which the spirit matures makes a huge contribution to the flavour of the mature whisky – in extreme cases determining over 80% of the flavour.

By law, Scotch whisky must be matured in oak casks. Most of the casks (around 90%) are made from American white oak (*Quercus alba*), the rest from European oak (*Quercus robur*).

The species of oak wood influences colour, aroma and taste. American oak imparts a golden hue and sweet, vanilla or coconut flavours; European oak, being more tannic, makes for a richer, umbrageous colour, a drier taste and typically introduces dried fruits, nuts and spices to the flavour, and sometimes sulphury notes.

Only very rarely are the casks brand new; the vast majority are seasoned prior to filling with Scotch spirit, either with American bourbon or Spanish sherry. Ex-bourbon casks are always made from American oak, and seasoned for three years; ex-sherry casks (usually Oloroso sherry) may be Spanish or American oak, and are seasoned for between one and four years. Occasionally ex-wine barrels, or wine-treated casks are used (these are usually European oak).

Casks come in a range of sizes. The commonest are 'barrels' (*c.* 200 litres, also called 'American Standard Barrels' or 'A.S.B.s'), 'hogsheads' (250 litres, also called 'remade hogsheads' – they are made by combining staves from A.S.B.s; four of the latter make three hogsheads), 'butts' (500 litres, usually made from European oak and ex-sherry) and 'puncheons' (also 500 litres, dumpier than butts; made from both European and American oak).

Casks are used three or four times before they are deemed to be 'exhausted'. The first time they are filled with Scotch they are termed 'first-fill casks', thereafter they are 'refill casks'.

Once they have become exhausted, they may be 'rejuvenated' by being scraped out and retoasted or charred.

Warehouses

Warehouses are of three kinds: traditional (or dunnage), racked and palletised.

Dunnage warehouses are low, damp and cool. The casks rest on top of each other, three high.

Racked warehouses are higher and drier. They are fitted with metal frames, into which rows of casks may be mechanically slotted, up to around 14 high.

Palletised warehouses are a recent innovation. Here the casks stand upright – not on their sides, as in the other two – six to a wooden pallet, which is then mechanically raised into position, as in a racked warehouse.

As the spirit matures it 'breathes' through the walls of the cask. Harsh alcohols are expelled – 2% by volume per annum is allowed by Customs & Excise, the so-called 'Angel's Share' – and water vapour seeps in.

In a cool, damp warehouse, the volume lost during maturation will be less than in a dry warm warehouse, but the strength will be lower (owing to dilution by incoming water vapour). In a dry, warm warehouse, the opposite will happen.

Most distilleries have warehouses on site, but few have enough space to accommodate all their casks. Accordingly, the spirit is shipped, usually by road tanker, to be filled into cask and matured elsewhere in Scotland. To be called 'Scotch whisky' the spirit, both malt and grain, must be matured in Scotland, and it does not become 'whisky' until it has matured for three years.

Individual warehouses, their location and microclimates make a contribution to flavour, but it is a small one in relation to other factors.

Style

The 'style' in each entry refers to the new-make spirit, not the mature whisky. The descriptors used here are mainly those used by the distillery owner.

Mature character

Wherever possible, my evaluation of this is based upon samples at around 12YO, bottled by the distillery owner (i.e. 'Proprietary Bottlings'). Clearly the character will change in older expressions (typically becoming more complex, deeper and often richer), single cask bottlings and wood-finished expressions (see 'Understanding the Label').

Aberfeldy

REGION
Highland (South)

ADDRESS
Aberfeldy, Perthshire

PHONE
01887 822010

WEBSITE
www.aberfeldy.com

OWNER
John Dewar & Sons Ltd
(Bacardi)

VISITORS
Dewar's World of Whisky
visitor centre

CAPACITY
3.5m L.P.A.

HISTORICAL NOTES: Aberfeldy distillery was built a quarter of a mile to the east of Aberfeldy village by the Dewar brothers, Tommy and John (Jnr), to supply fillings for their increasingly successful blends. The site chosen was that of the former Pitilie Brewery, which had also once been a distillery of the same name, operating from 1825 to 1867. The founders' father, John Dewar, was born on a croft about two miles away. The new distillery opened in 1898 and was built according to 'the most modern principles': barley went in one end of the distillery and whisky came out at the other. There was even a private railway line connecting with Dewar's blending operation in Perth, to bring in grain and coal, and to carry away casks of whisky. An old 'Saddle' shunting locomotive, known as the 'Puggie' (a Scots word for 'monkey'), at the distillery is all that is left of this railway connection, which closed during the 1960s.

The distillery passed to D.C.L., along with the brand, when Dewar's joined in 1925 (the 'Big Amalgamation'), and was subsequently operated by S.M.D. When building regulations were relaxed in 1960 S.M.D. replaced Aberfeldy's two stills with exact copies, and converted to mechanical coal-stoking; in 1972/3, the stillhouse and tun room were rebuilt using the original stone, now with four indirect fired stills.

When D.C.L.'s successor U.D. merged with Grand Metropolitan in 1998 to become U.D.V. (and later Diageo) the company was obliged to divest some of its spirits interests. The Dewar brands and four distilleries, including Aberfeldy, were bought by the Bacardi Corporation.

Since then, over £3 million has been spent on the distillery, including the creation of a first-rate visitor centre, *Dewar's World of Whisky*, which makes good use of the company's extensive archive to tell the story of the remarkable Dewar brothers and of the global success of the brand, which has long been

The company has only had seven master blenders in the past 100 years.

number one in the United States. It opened in 2000 and welcomes around 35,000 visitors a year.

CURIOSITIES: The ground upon which Aberfeldy stands was feued from the Marquis of Breadalbane, who retained the right to mine gold there. This right was acquired in 2007 by Alba Mineral Resources, who are currently embarked on research into potential gold reserves in a 322.4 sq. km. area around Loch Tay.

Dewar's master blender in 1902 was A.J. Cameron, who pioneered the idea of returning whisky to cask after blending to allow the spirits to marry for three to six months. His successor, Stephanie Macleod, continues this for Aberfeldy single malt, as well as for the blends. (Dewar's Signature blend is married for two years.)

In 2007 Bacardi bought land at Poneil, near Motherwell, where they have built nine new maturation warehouses, with a further nine to come, at a total cost of around £120 million.

RAW MATERIALS: Unpeated malt from Simpson's, Berwick-upon-Tweed; water from the Pitilie Burn.

PLANT: Full-Lauter mash tun (6.3 tonnes); eight Siberian larch washbacks; two of stainless steel. Two plain wash stills (charge 16,500 litres); two plain spirit stills (charge 15,000 litres). All indirect fired by steam coils. Shell-and-tube condensers.

MATURATION: Around 90% refill U.S. hogsheads, 10% refill European wood. All Dewar's spirit filled and matured in Glasgow.

STYLE: Sweet and estery; heather honey and Ogen melon.

MATURE CHARACTER: Smooth and creamy; honeycomb; pears, melon and bruised apples; light maltiness. Taste is fresh, fruity and waxy, predominantly sweet. Medium body.

Aberlour

REGION
Speyside
ADDRESS
Aberlour, Moray
PHONE
01340 881249
WEBSITE
www.aberlour.com
OWNER
Chivas Brothers
VISITORS
An elegant new visitor
centre opened in
August 2002

3.8m L.P.A.

HISTORICAL NOTES: A distillery was founded in the grounds of Aberlour House in 1825 by the laird, James Gordon, and a man named Peter Weir. By 1833 this venture had failed, and the distillery had been taken over by John and James Grant, in partnership with two brothers named Walker. When the lease ended in 1840, the Grants moved to Rothes and built Glen Grant Distillery – Peter Weir's son became their commercial traveller – and the Walkers went to Linkwood Distillery.

Today's Aberlour Distillery was built in 1879/80 by James Fleming (a local businessman who leased Dailuaine Distillery until that year), a mile from the original, using stone from the quarry used by Thomas Telford for his famous bridge at Craigellachie. It was enlarged in 1892, when ownership passed to the Greenock blenders, R. Thorne & Sons. It was largely destroyed by fire in 1898 and rebuilt.

The next owner was the English company W.H. Holt & Sons (Chorlton-cum-Hardy) Ltd, who bought Aberlour in 1921 and ran it for 20 years. In 1945 it was bought by S. Campbell & Son and re-equipped in 1973 (four stills), then sold to Société Pernod Ricard the following year. Pernod retained Campbell Distillers as its whisky division until acquiring Seagram's in 2001, including Chivas Brothers Ltd, which took over management of all its distilleries.

Sales of Aberlour single malt have increased dramatically since 2000, particularly in France, where it is the bestseller – 3.5 million bottles were sold in 2014.

A new visitor centre was opened in August 2002, offering an excellent two-hour tour.

CURIOSITIES: Until the 1890s the distillery was entirely powered by water. In the grounds there is a holy well dedicated to Saint Drostan – a follower of Saint Columba who visited Speyside around

The name means 'the mouth of the chattering burn'.

660AD – reputedly used by him for baptising wild Highlanders. He later established a monastery at Deer, Aberdeenshire, where the famous *Book of Deer* was later written. A small church dedicated to this saint opened in Rothes in 1931.

Ian Mitchell, who died in 1992, worked at the distillery for 48 years, the last 27 of them as Manager. His grandfather, father and brother also worked there.

Aberlour a'bunadh ('original' in Gaelic) was introduced in 2000 – a natural strength, un-chill filtered expression from ex-sherry casks. To 2014 there have been 47 batches of this excellent single malt.

Aberlour House became a prep school, and it is now the headquarters of Walkers of Aberlour, the famous shortbread bakers.

RAW MATERIALS: Unpeated malt from independent maltsters; own floor maltings closed in 1960. Process and reduction water from springs; cooling water from the Lour Burn.

PLANT: Semi-Lauter mash tun (12.12 tonnes); six stainless steel washbacks. Two plain wash stills (charge 12,500 each); two plain spirit stills (charge 16,000 litres each, requiring two wash distillations to charge one still). All indirect fired by steam. External shell-and-tube condensers.

MATURATION: Approximately half is first and refill sherry butts, the remainder refill hogsheads. The whisky to be bottled as single is matured on site in one of six bonded warehouses (capacity 27,000 casks); the rest at other sites in the Highlands and Central Belt.

STYLE: Sweet and fresh-fruity, medium bodied.

MATURE CHARACTER: Malty nose, with some fruits and spice. Viscous mouthfeel. Honey, with a trace of nutmeg and a thread of smoke in the finish. Medium-bodied.

Abhainn Dearg

REGION
Island (Lewis)

ADDRESS
Carnish, Nr Uig, Isle of
Lewis

PHONE
01851 672429

WEBSITE
www.abhainndearg.co.uk

OWNER
Mark Tayburn

VISITORS
Welcome

CAPACITY
c. 30,000 L.P.A,

HISTORICAL NOTES: Abhainn Dearg Distillery (pronounced 'Aveen Jerrak', meaning 'Red River') is currently the smallest in Scotland, and also the most westerly, being situated at Uig on the Atlantic coast of Lewis.

It is the creation of Mark Tayburn, a Lewisach born and bred, and commenced production in September 2008. Part of the distillery makes use of a former salmon hatchery (some of the hatchery pens are still in use, producing brown trout and char to be released in nearby Loch Scaslavat); the still-house itself is new and utilitarian.

The distillery is proudly artisanal. The curious still designs, with crazily tilted still heads – as conical as witches' hats and with spidery lyne arms – is based on an illicit still found on the island in the 1950s (and displayed at Abhainn Dearg) and Mark's intention is that in due course all his barley will be sourced locally and malted on site – he is already growing Golden Promise, and fattening his Highland cattle on draff.

CURIOSITIES: The 'Red River' is named on account of a bloody skirmish over a thousand years ago, when the people of Lewis drove off a band of marauding Vikings with so many casualties that the lower reaches of the Abhainn Caslabhat ran red with blood.

Martin Martin, who published his *Description of the Western Isles* in 1703, wrote of Lewis:

'Their plenty of Corn was such as disposed the Natives to brew several sorts of Liquors, as common *Usquebaugh*, another call'd *Trestarig*, *id est*, *Aqua-vitae*, three times distilled, which is strong and hot; a third sort is four-times distill'd, and this by the Natives is call'd *Usquebaugh-baul* [literally 'Perilous whisky'], *id est Usquebaugh*, which at first taste affects all the Members of the Body: two spoonfuls of this Last Liquor is a sufficient dose; and if any man exceed this, it would presently stop his Breath, and endanger his Life.'

'The Collector of the Customs at Stornoway, in the Isle of Lewis, told me that about one hundred and twenty families drink yearly four thousand English gallons of this spirit [i.e. *uisge beatha*] and Brandy.' (Capt. Edward Burt, *Letters From A Gentleman in the North of Scotland*, 1754)

The only previous licensed distillery on Lewis was Shoeburn, in Stornoway (*c.* 1830–1840). Soon after it closed, the island was bought by Sir James Matheson, a total abstainer (see also Dalomore); he demolished the buildings in Shoeburn Gorge and incorporated the site into Lews Castle. Whisky production continued, but underground.

The first bottling of Abhainn Dearg was released in September 2011 at the Royal National Gaelic Mod, which was held at Stornoway that year. It is named *Spirit of Lewis*.

Some new-make spirit has also been bottled and is for sale at the distillery.

RAW MATERIALS: Soft, mineral-rich water from Loch Raonasgail, via the Abhainn Caslabhat and the Abhainn Dearg. Peated malt currently from independent maltsters in northern Scotland. Plans to install own maltings and use locally grown barley.

PLANT: Modelled on a former illicit still, there are two mash tuns (500kg each) and two Douglas fir washbacks. Both wash still (capacity 2,112 litres) and spirit still (2,057 litres) have high conical necks and steeply descending lye pipes, both with worms in wooden tubs.

MATURATION: Ex-bourbon barrels, with plans to fill some ex-sherry casks.

STYLE: Rich and peaty.

Ailsa Bay

REGION
Lowland

ADDRESS
Girvan, Ayrshire

PHONE
01465 713091

WEBSITE
No

OWNER
William Grant & Sons Ltd

VISITORS
No

CAPACITY
12m L.P.A.

Ailsa Bay is the third largest malt whisky distillery.

HISTORICAL NOTES: William Grant & Sons announced their intention to build a new distillery within their Girvan grain spirit distillery in January 2007. By December the same year, spirit samples were being sent to potential customers. The project reputedly cost in excess of £10 million.

CURIOSITIES: 'The most outstanding natural feature of the area is the uninhabited island of Ailsa Craig, standing a few miles off the coast. It is the granite plug of a very old volcano. The granite from this extraordinary, steep-sided island is known as "ailsite", and until recently was the sole source of curling stones for Scotland's national winter sport.' (Philip Morrice, *The Whisky Distilleries of Scotland and Ireland*, 1987)

This is the second pot-still malt distillery to have been built within William Grant's Girvan site. The first was Ladyburn (see entry), which operated from 1966 to 1975. The new distillery does not have exactly the same location as the old one, nor does it employ any of its equipment.

The make is used for Grant's blends, and is not yet available as a single malt.

RAW MATERIALS: Unpeated malt from independent maltsters. Soft water from Penwhapple Loch.

PLANT: Full-Lauter mash tun (12 tonnes); 12 stainless steel washbacks. The stills have been modelled on those at Balvenie: four straight-side boil-ball wash stills (12,000 litres charge); four boil-ball spirit stills (12,000 litres charge). All indirect-fired. Shell-and-tube condensers.

MATURATION: Mix of first-fill and refill ex-bourbon; small amount of European oak casks.

STYLE: Rich, complex, sweet. Balvenie style.

Allt–a–Bhainne

REGION

Speyside

ADDRESS

Glenrinnes, by Dufftown, Moray

PHONE

01343 554120

WEBSITE

No

OWNER

Chivas Brothers

VISITORS

No

CAPACITY

4m L.P.A.

The distillery looks out over to Corriehabbie, the haunt of many illicit distillers in the past, whose product was called Leprach, *after one of the neighbouring hills.*

HISTORICAL NOTES: Built by Seagram's in 1975, Allt-a-Bhainne Distillery looks like a compact fortress crouching on the northern slopes of Ben Rinnes, the mountain that dominates northern Speyside.

The distillery itself is uncompromisingly modern, inside and out, but the six blocks that make up the building are cleverly heaped together, each with a different roofline, and the most forward block is clad with local stone, so that the overall effect is pleasing to the eye.

Production commenced with one pair of stills; a further pair were added in 1989. The whole operation was designed to be operated by one man, even in pre-computer days. The whisky made here is all carried to Keith by tanker and filled into cask there.

CURIOSITIES: Two years before Allt-a-Bhainne was commissioned, Seagram's had built Braeval as a result of the huge demand for Scotch at the time and anticipating that it would continue. Alas, this was not to be the case, and both distilleries have endured periods of closure.

Never bottled by its owner. Very occasional independent bottlings only.

RAW MATERIALS: Unpeated malt from independent maltsters. Unusually, the distillery has a Buhler five-roller malt mill. Water from 20 springs on the sides of Ben Rinnes, via the Rowantree and Scurran Burns, is collected in a dam behind the distillery.

PLANT: Traditional rake-and-plough mash tun (nine tonnes); eight stainless steel washbacks.

Two plain wash stills (22,000 litres charge); two tall boil-ball spirit stills (22,000 litres charge). The original spirit still has a straight neck, and was adapted in the 1980s. All steam fired. Shell-and-tube condensers.

MATURATION: Mainly refill U.S. hogsheads, matured off site.

STYLE: Sweet and grassy.

Annandale

REGION
Lowland

ADDRESS
Northfield, Annan,
Dumfriesshire

PHONE
01461 207817

WEBSITE
www.annandaledistillery.
co.uk

OWNER
David Thomson and
Teresa Church

VISITORS
Visitor centre

CAPACITY
500,000 L.P.A.

HISTORICAL NOTES: The first Annandale Distillery was founded in 1835 and operated, latterly under the ownership of John Walker & Sons, until 1920. The site was bought by Professor David Thomson and his wife, Teresa Church, in 2007 and, after painstaking archaeological survey, they have completed extensive restorations to create 'a modern distillery which rises sympathetically on the site of its forbear'.

Work on the restoration began in June 2011 with the repair of two old sandstone warehouses. A new mill room housing four 15-tonne malt bins and Porteous mill (ex-Caperdonich) has been created within the existing maltings. An exact copy of the mash house has been rebuilt on the original footprint which now houses the mash tun and six washbacks. A new stillhouse has been created in what was formerly the mill house. The unique visitor centre has been created in the former maltings, designed and built by craft furniture maker, Ian Cameron-Smith. The original Charles Doig-designed pagoda kiln roof has been carefully restored. The distilling plant has been made by Forsyth of Rothes, and Malcolm Rennie, formerly of Kilchoman Distillery, has been appointed manager.

Annandale is unusual in having two low wines and feints receivers, allowing production to switch readily between smoky/peaty and non-smoky/peaty expressions. It is also quite unusual in having one large wash still serving two spirit stills.

RAW MATERIALS: Peated and unpeated malt, and a combination of Mauri and dried yeast. Process water is drawn from the distillery's own bore hole.

PLANT: Stainless steel semi-Lauter mash tun (2.5 tonnes), six 12,000-litre Douglas fir washbacks, one 12,000-litre wash still and two 4,000-litre spirit stills with boil ball. External shell-and-tube condensers with adiabatic cooling.

MATURATION: Combination of first and refill ex-sherry and ex-American whiskey casks, on site.

STYLE: There will be two expressions: moderately smoky/peaty with distinct sherry character and a non-peaty, fruity expression with minimal bourbon character. It is intended that almost all of the production will be sold by the distillery as single malts but some casks will be made available to specialist bottlers and private purchasers.

Arbikie

Ar

REGION
Highlands (East)
ADDRESS
Inverkeillor, Arbroath, Angus
PHONE
01241 830770
WEBSITE
www.arbikie.com
OWNER
The Stirling Brothers
VISITORS
No
CAPACITY
200,000 L.P.A.

HISTORICAL NOTES: Arbikie Highland Estate is a 2,000-acre farm overlooking the white sands of Lunan Bay with the Angus hills to the west. It has been owned by the Stirling family for four generations. The rich red earth of Angus is among the most fertile in Scotland, and there are records from 1794 of distilling on the site. Between 2013 and 2014 three Stirling brothers, Iain, John and David, built an artisanal distillery with both pot and column stills from CARL (the oldest still fabricators in Germany) to produce malt whisky, gin and vodka. All the spirits are made initially in the pot stills, with the vodka (made from potatoes, which is also the base material for the gin) then being rectified in the column stills.

It is a 'single-site operation': the raw materials (barley and potatoes) are grown on the farm, water comes from hills to the west and fills an underground lagoon, and bottling and labelling is done within the distillery building itself, which is housed in a refurbished cattle shed.

Kirsty Black and Christian Perez, both graduates in brewing and distilling from Heriot Watt University, run the distillery and all the production within it. Arbikie launched Scotland's first potato vodka in 2014, an 'ultra-smooth' vodka, made from Maris Piper, King Edward and Cultra potatoes; subsequently Arbikie launched Kirsty's Gin, using locally sourced botanicals. At the end of 2015 Arbikie launched a chilli vodka, adding chipotle chillis (grown in Fife) to the same base.

CURIOSITIES: Arbikie is determined to continue to grow all its ingredients for producing alcohol. Expect a number of 'seasonal' releases that reflect the growing seasons in Scotland.

A downward-angled lyne arm combined with an optional purifier on the spirit still gives the possibility of making a range of new spirit styles.

RAW MATERIALS: Barley grown on the estate and malted by Boortmalt UK, Hillside – less than ten miles away from the distillery. Soft water from springs and burns in the Angus hills.

PLANT: Stainless steel semi-Lauter tun (0.75 tonnes). Four stainless steel washbacks. One CARL wash still (4,000 litres charge), one CARL spirit still (2,400 litres charge), the latter connected to a CARL column still, all indirect fired. Shell-and-tube condensers.

MATURATION: A mix of casks including bourbon, sherry and wine.

STYLE: A Highland coastal spirit with a rich deep flavour and honeyed tones.

Ardbeg

REGION
Islay
ADDRESS
Port Ellen, Isle of Islay,
Argyll
PHONE
01496 302244
WEBSITE
www.ardbeg.com
OWNER
The Glenmorangie
Company
VISITORS
Centre, stylish shop and
excellent café/restaurant
CAPACITY
1.3m L.P.A

HISTORICAL NOTES: Ardbeg has an old-fashioned, timeless atmosphere, with tasteful contemporary touches. It was resurrected from the dead by Glenmorangie, after they bought the distillery in 1997 for £7 million – over £6 million of which was for stock, the buildings and plant were so dilapidated.

The founders of the distillery were MacDougalls, tenant farmers in this part of Islay, and this family retained connections with Ardbeg until the 1970s. The first record of distilling on the site was in the 1790s, but the earliest commercial operation dates from 1815. By 1900 the village around the distillery housed the families of 40 workers (and two Excise officers); the school had over 100 pupils. However, it was in decline by the end of the 1920s. Ardbeg single malt was available by the small cask to private customers 'who had a credit account with the distillery'.

Alexander MacDougall & Company continued to own the distillery until the firm's liquidation in 1959, when Ardbeg Distillery Ltd was formed. This became the Ardbeg Distillery Trust in 1973 when the distillery was taken over by a joint venture between D.C.L. and Hiram Walker & Sons, the Canadian distiller. The latter acquired full control in 1976 (paying £300,000), but the distillery was mothballed five years later, with the loss of 18 jobs. This was the end of Ardbeg village.

Hiram Walker's distilling interests were bought by Allied Lyons (later Allied Distillers) in 1987 and production resumed in 1989 on a small scale. But Allied also owned Laphroaig Distillery, just down the road and producing a similarly smoky malt, so Ardbeg was again mothballed in 1996 and discreetly offered for sale. The successful bidder was Glenmorangie, which took over in 1997 and spent £1.4m on restoration, new plant and a visitor centre. Production resumed the following year under the management of Stuart Thomson (until

The Ardbeg Committee, a fan club founded in 2000, now has around 56,000 members in 120 countries.

2007), assisted by his wife, Jackie, who created the excellent visitor centre.

CURIOSITIES: Alfred Barnard mentions the clan loyalty of the MacDougalls. When Alexander MacDougall of Ardbeg discovered that a kinsman had been found guilty of some misdemeanour, he immediately paid the fine, saying 'that it was impossible that a MacDougall could do anything wrong'.

Ardbeg was (and is) famously smoky whisky. Until 1977, it used its own kilns for drying malt, and it is said that the louvers in the kilns were manipulated so as to cause the smoke to linger and impart peatiness at 50ppm phenols. Since this date, malt has come from Port Ellen Maltings, specified at the same level. In spite of this high phenol level, the whisky itself does not taste as smoky as some other Islays, and this may be owing to the purifier pipe on the spirit still, which returns any liquid which has condensed in the lyne arm for redistillation.

Until the roll-on-roll-off ferry service to Islay was launched in 1968, Ardbeg's barley and coal arrived by sea, and its casks of whisky went out the same way. It was not always plain sailing. In December 1925 the puffer *Serb* was wrecked on the rocks at the entry to Ardbeg Bay, with a cargo of barley and malt. The crew were saved.

For two years after 1979 and between 1989 and 1996 Ardbeg produced an unpeated malt for blending purposes, called Kildalton – the equivalent of Caol Ila's 'Highland' style. A 1980 expression was bottled in 2004.

RAW MATERIALS: Soft peaty water from Loch Arinambeist and Loch Uigeadail, three miles from distillery. Floor maltings until 1977 then heavily peated malt from Port Ellen Maltings (55ppm phenols).

PLANT: Stainless steel semi-Lauter mash tun (4.5 tonnes). Three larch washbacks and three of Douglas fir. One lamp-glass wash still (11,775 litres charge) and one lamp-glass spirit still (13,660 litres charge), both indirect fired by steam. Shell-and-tube condensers.

MATURATION: 98% bourbon barrels (first and second fill) and 2% sherry. 50% first-fill and 50% second-fill used for single bottlings. Two dunnage and three racked warehouses on site (capacity 24,000 casks).

STYLE: Peaty.

MATURE CHARACTER: Peaty, medicinal, salty, dry, but the taste is surprisingly sweet, followed by a blast of smoke, with some liquorice. Full-bodied.

Ardmore

REGION
Highland (East)

ADDRESS
Kennethmont, by Huntly,
Aberdeenshire

PHONE
01464 831213

WEBSITE
www.ardmorewhisky.com

OWNER
Suntory

VISITORS
By appointment

CAPACITY
5.53m L.P.A.

HISTORICAL NOTES: Ardmore is the heart malt of the Teacher's blends, and was built by Adam Teacher (son of the founder, who died shortly before the distillery was completed) in 1897/8. It was the family's first venture into distilling – in order to secure spirit supply for their successful Highland Cream brand (introduced 1884). The company would later buy Glendronach Distillery (1960).

The distillery stands deep in rural Aberdeenshire, between the historic villages of Spynie and Kennethmont, beside the main Inverness–Aberdeen railway line, off which the distillery had a siding (no longer used). Capacity was doubled to four stills in 1955, and eight stills in 1974 – all were direct fired by coal until 2002. The original steam engine, used to provide power, is still in situ and operational. Until 1976 malting in Saladin boxes was done on site.

William Teacher & Sons was acquired by Allied Breweries in 1976. The owner never bottled Ardmore as a single malt, and only a small amount found its way to Gordon & Macphail and Cadenhead's. When Allied Domecq was sold in 2006, Teacher's brands, together with Ardmore and Laphroaig Distilleries, were sold to the American distiller Beam Global (owner of Jim Beam bourbon and a subsidiary of Fortune Brands). The company's Scotch Whisky Director, Douglas Reid, whose father worked at Ardmore for many years, and who was raised in one of the distillery cottages, plans to pursue a policy that 'provides continuity and builds on local knowledge, skill and experience gained over many years and generations'. How wise!

In March 2014 it was announced that the Japanese distillery Suntory (owner of Bowmore, etc.) had bought Beam Inc. for £9.8 billion, including Teacher's, Ardmore and Laphroaig, Courvoisier Cognac, Jim Beam and Maker's Mark bourbons and Canadian Club.

CURIOSITIES: Unusually smoky for a Highland malt, with a peating specification around 12–14ppm phenols. The peat comes from Saint Fergus in Buchan and has a

Teacher's Highland Cream was the first whisky to use a stopper cork, introduced in 1913. The firm advertised under the slogan 'Bury the Corkscrew', and Teacher's was described as 'The Self-Opening Bottle (Patented)'.

different character to west coast peat, imparting a dry earthiness to the flavour of the whisky.

Before the invention of the stopper cork by William Manera Bergius, Adam Teacher's nephew, all whisky bottles had driven corks, like today's wine bottles.

Ardmore was one of the last distilleries to fire its stills directly with coal. This was only abandoned in 2002, in favour of indirect firing by steam-heated coils and pans.

Ardmore and its sister, Laphroaig, pioneered the use of quarter cask finishing in first-fill American oak barrels of around 100-litres capacity.

The only expression of Ardmore currently bottled by its owner is Traditional Cask, with no age statement, @ 46% vol without chill-filtration, finished in the aforementioned quarter casks.

RAW MATERIALS: Soft process water from 15 springs on Knockandy Hill. Cooling water from local sources. Medium-peated malt (12–14ppm phenols) from local independent maltsters.

PLANT: Traditional cast iron mash tun, semi-Lauter stirring gear, copper dome (12.5 tonnes). Fourteen Douglas fir washbacks. Four plain wash stills (15,000 litres charge each); four plain spirit stills, but narrower and slightly taller than the wash stills (15,500 litres charge each). All indirect fired since 2002. Shell-and-tube condensers with after-coolers.

MATURATION: The majority of the spirit used for Teacher's is matured in European oak puncheons. The single malt is matured in ex-bourbon (Jim Beam) barrels, and finished in first-fill quarter casks.

STYLE: Sweet and smoky, with spice in the finish.

MATURE CHARACTER: Creamy, sweet and smoky on the nose; mellow and buttery, sweet and malty; distinctly smoky to taste. Unusual. Robust.

Ardnamurchan

REGION
Highland (West)

ADDRESS
Glenbeg, Ardnamurchan,
Argyll

PHONE
01972 500285

WEBSITE
www.adelphidistillery.com

OWNER
Adelphi Distillery Ltd

VISITORS
Visitor centre

CAPACITY
500,000 L.P.A.

HISTORICAL NOTES: Probably Scotland's most remote mainland distillery, Ardnamurchan commenced production in June 2014. It has been built by the highly regarded independent bottler Adelphi, and takes its name from its location: the Ardnamurchan peninsula, which is the most westerly point of the British mainland and is owned by one of Adelphi's directors, Donald Houston.

The distillery commands splendid views across Loch Sunart to the Isle of Mull, and is unique in Scotland in being entirely powered by hydro-energy and a large bio-mass boiler, fuelled by timber from Ardnamurchan Estate. This was the brainchild of Mr Houston, who is a highly skilled and successful engineer. The boiler will also provide heat for the distillery's floor maltings – another unusual feature.

The distillery's first manager was Graeme Bowie, with thirty years' experience in the industry, at Glen Grant and Balmenach, and latterly as assistant manager at Balblair Distillery.

CURIOSITIES: The original Adelphi Distillery was built in 1826 by brothers Charles and David Gray, on the banks of the River Clyde in what is now the heart of Glasgow. The two-acre site had been an orchard, fronted by a wharf, and stood just south of the Clyde's Victoria Bridge on the northern edge of the Gorbals.

Around 1880 it was taken over by Messrs A. Walker & Co., owners of distilleries in Limerick and Liverpool – the latter was the largest distillery in the U.K. at the time – who invested heavily and expanded and improved the business until it became one of the most advanced and productive distilleries in Scotland.

In 1993, Jamie Walker (the great-grandson of Archibald Walker) revived the Adelphi name as an independent bottler. He sold the company to Donald Houston and his neighbour in Argyll, Keith Falconer of Laudale Estate. The company's managing director, Alex Bruce, is the son of the Earl of Elgin and Kincardine, whose estate in Fife supplies the barley for the distillery. Adelphi's HQ and bottling facility is also located here.

RAW MATERIALS: Process water from springs above the distillery; cooling water from the Glenmore River. Barley from Broomhall Estate in Fife, 40% malted without peating; 60% malted on site, heavily peated.

PLANT: Stainless steel semi-Lauter mash tun (2 tonnes). Four oak washbacks (unique; ex-cognac vats); three stainless steel washbacks. One plain wash still (10,000 litres), one lamp-glass spirit still (6,000 litres charge). External shell-and-tube condensers.

MATURATION: In first and refill ex-bourbon and ex-sherry casks, on site.

STYLE: The owners plan to distil two styles of spirit: unpeated and heavily peated. These will be matured separately then blended to produce a 'West Highland' style of whisky. Occasional single casks of heavily peated malt will be released under the Adelphi label.

Arran

REGION
Highland (Island)

ADDRESS
Lochranza, Isle of Arran,
North Ayrshire

PHONE
01770 830264

WEBSITE
www.arranwhisky.com

OWNER
Isle of Arran Distillers Ltd

VISITORS
Visitor centre, A/V presen-
tation and restaurant

CAPACITY
1.5m L.P.A.

HISTORICAL NOTES: The Isle of Arran Distillery was the brain-child of Harold Currie, former Managing Director of Chivas Brothers and later of Campbell Distillers. Some money was raised by a novel 'bond-holder' scheme, which invited subscribers to invest by guaranteeing them a certain amount of whisky – five cases of blended whisky in 1998, five cases of Arran Founder's Reserve in 2001, all for £450. The distillery, which stands above the picturesque village and sea loch of Lochranza, opened in 1995.

When the distillery opened, there was much speculation about which style of malt it would produce. Would it be Islay smoky, Campbeltown heavy or Lowland light? The distillery stands at the crossroads. Even by 1995, however, the influence of the location could largely be controlled by chemistry. Harold Currie was a Speyside distiller, and opted for that style. So it is impossible to classify Arran by regional style. Mr Currie died on 15 March 2016, aged 91. A week later, the company announced plans to build another distillery at Lagg in the south of the island.

CURIOSITIES: The whiskies of Arran were rivalled only by those of Glenlivet by one nineteenth-century source, although there was only ever one licensed distillery on the island (at Lagg, in the southern part, closed 1837).

The visitor centre, which later won a top award from VisitScotland, was opened by H.M. The Queen in 1997 and is one of the island's main visitor attractions. In 2015 it welcomed 88,000 visitors.

As well as its 'Arran Malts' (unpeated) and 'Machrie Moor' malts (peated), the distillery also markets a range of blends and malts under the Robert Burns label, officially recognised by the International Burns Club network, of which the distillery is a patron. Unlike most distilleries,

Golden eagles are a familiar sight above the distillery, which is encircled by the high mountains of the north of Arran.

you may buy the Arran malt by the cask, as new-make spirit.

RAW MATERIALS: Water from Loch na Davie, above the distillery via the Easan Biorach stream. 90% unpeated Scottish malt from Bairds (Black Isle). Each year a small amount of peated malt is taken in, for a peated expression.

PLANT: Stainless steel full-Lauter mash tun (2.5 tonnes). Four Oregon pine washbacks. One plain wash still (6,500 litres charge) and one plain spirit still (3,695 litres charge), both indirect fired by steam coils. Shell-and-tube condensers.

MATURATION: Two modern dunnage warehouses, one racked on site, holding 5,000 casks – 80% of the make. The remaining 20% is matured on the mainland.

STYLE: Sweet and fruity, eau-de-vie.

MATURE CHARACTER: Speyside-like; pear drops, citric fruits, green apples. The taste is sweet, with some malt and citric acidity. Light-bodied.

Auchentoshan

REGION
Lowland
ADDRESS
Dalmuir, Dunbartonshire
PHONE
01389 878561
WEBSITE
www.auchentoshan.com
OWNER
Beam Suntory
VISITORS
Visitor centre, conference
facilities, shop
CAPACITY
2.1m L.P.A.

HISTORICAL NOTES: Triple distillation was common among Lowland distilleries, many of which distilled other grains than malted barley in their pot stills. It increases the delicacy of the whisky, and its strength. Auchentoshan is the sole remaining Lowland distillery employing triple distillation, and until 2014 was one of only four surviving malt distilleries in the Lowlands.

It was founded by John Bulloch, corn merchant, in 1817 (as Duntocher), but he filed for bankruptcy in 1822. His son, Archibald, obtained a licence under the 1823 Act, but also went bust in 1826. In 1834 Duntocher was sold to John Hart (distiller for Bulloch & Co.) and Alexander Filshie, a local farmer whose family had lived in the district since the seventeenth century, on condition that John Bulloch be allowed to continue to live on the distillery grounds. (He died in 1846, aged 87.)

They changed its name to Auchintoshan (sic), which means 'the corner of the field'. Members of the Filshie family owned the distillery for 44 years. They rebuilt it in 1875 then sold it to C.H. Curtis & Company, whisky merchants in Greenock, following a disastrous grain harvest in 1877.

Curtis & Co. ran Auchentoshan until 1900, when it again changed hands, twice in three years; the second time going to John and George McLachlan Ltd, 'Brewers, Distillers and Wine & Spirits Merchants', who sold to Tennent's (Brewers) in 1960. Tennent's became part of Charrington in 1964 (later Bass Charrington), and in 1969 the distillery was sold to Eadie Cairns (for around £100,000). They re-equipped and refurbished, and sold it to its present owners in 1984. The Japanese distiller Suntory bought Morrison Bowmore in 1994, and Suntory merged with Jim Beam in 2015.

CURIOSITIES: Auchentoshan's situation 20 minutes west of Glasgow makes it a popular place to visit, and the facilities there were refurbished in 2004

In 1941 Auchentoshan was badly damaged by enemy action, which destroyed three warehouses and almost one million litres of maturing spirit. Clydebank was levelled and 1,000 people killed, but miraculously the distillery survived. Today cooling water is collected from a bomb crater near the distillery. Production was resumed in 1948.

and 2008 (with 5* accreditation from VisitScotland) to enhance the experience. Its entire output is bottled as single malt and promoted as 'The Spirit of Glasgow'.

The Bulloch family went on to play an important role in the whisky industry. Soon after they disposed of Auchentoshan, Bulloch & Company was founded (1830), and in 1855 John Bulloch's grandsons merged with the successful Glasgow whisky merchants A. Lade & Company to become Bulloch Lade & Company. Camlachie/Loch Katrine distillery was acquired next year, Caol Ila in 1863 and Benmore (Campbeltown) in 1868. The firm also offered a range of well known blends, under the generic title 'B.L.', and by 1893 were sole agents for the Macallan-Glenlivet Distillery. Incorporated in 1898, Bulloch Lade joined D.C.L. in 1920.

Triple distillation works as follows: wash charges the wash still, and the low wines from here go to Feints Receiver 1, where they mix with the tails of the intermediate still (overall strength 18–19%). The smaller intermediate still is charged from this receiver, and the heads go to Feints Receiver 2, where they join the feints from the spirit still (overall strength 55%). The spirit still is charged from here, and distilled in the usual way, with foreshots and feints being returned to Feints Receiver 2, and spirit saved in the spirit receiver. Overall strength is an unusually high (for malt distilling) 82%.

A major revamp of packaging and liquids took place in 2008. Three Wood has featured in the finals of the Cuban Whisky and Cigar Challenge since its inception in 2003, and won in 2005 (paired with a Bolivar Imensas cigar).

RAW MATERIALS: Soft process water from Loch Katrine and cooling water from the Kilpatrick Hills, via the bomb crater dam where it is recyled through a fountain. Unpeated malt from independent maltsters.

PLANT: Stainless steel semi-Lauter mash tun, with copper canopy (6.82 tonnes). Four Douglas fir washbacks. One lamp-glass wash still (17,300 litres charge), one lamp-glass intermediate still (8,000 litres charge), one lamp-glass spirit still (11–12,000 litres charge). All indirect fired. External shell-and-tube condensers; those on the wash still are unusually large.

MATURATION: Ex-bourbon barrels, ex-sherry casks (Oloroso and Pedro-Jimenez), remade hogsheads and butts. Each expression has its own mix of casks/wood formula. Three dunnage and two racked warehouses on site hold 20,000 casks.

STYLE: Delicate, fruity and zesty.

MATURE CHARACTER: Floral-fragrant, with lemon zest and light cereal. Smooth mouthfeel, sweet then dry with roast almonds, fruits and a trace of butterscotch. Short finish. Light-bodied.

Auchroisk

REGION
Speyside
ADDRESS
Mulben, Moray
PHONE
01542 885000
WEBSITE
www.malts.com
OWNER
Diageo plc
VISITORS
By appointment
CAPACITY
5.9m L.P.A.

HISTORICAL NOTES: Commissioned by Justerini & Brooks (J. & B.), building started at Auchroisk in 1972, on a site to the west of Keith with quantities of high-quality water available from Dorie's Well. The name comes from a nearby farm and means 'the ford of the red stream'. The distillery was designed by Westminster Design Associates and built by George Wimpey & Co.; it was completed by 1974.

The building won an award from the Angling Foundation for not interfering with the salmon's progress up the Mulben Burn.

A year before building started, water was tankered over to Glen Spey Distillery where a week's trial was conducted – the commissioners wanted a malt with a light style similar to Glen Spey for their leading blend, J. & B. Rare. J. & B. had been part of Independent Distillers and Vintners (together with W. & A. Gilbey & Sons) since 1962. In 1972, while Auchroisk was being built, I.D.V. was bought by the brewer Watney Mann, and both companies were taken over by Grand Metropolitan within the year. Grand Metropolitan merged with Guinness to form United Distillers and Vintners, then Diageo, in 1997.

Auchroisk is a large site and casks from other distilleries are matured here. It also acts as Diageo's collection centre in the north of Scotland: casks of mature whisky arrive here, are disgorged, blended and filled into road tankers for transportation to blending halls in the Central Belt.

CURIOSITIES: Jim Milne, J. & B.'s Master Blender, took the unusual step of reracking The Singleton into ex-sherry casks for two years after ten years' maturation in refill ex-bourbon. This was possibly the first example of 'wood finishing', although it was not stated on the label. These early bottlings covered themselves in international awards.

The Singleton brand name was dropped for Auchroisk in 2001, but was revived in 2006 for

The Singleton brand name was adopted because the distillery name was deemed difficult to pronounce.

expressions of Glen Ord, Glendullan and Dufftown (see individual entries).

RAW MATERIALS: Unpeated malt from Burghead. Soft process water from Dorie's Well; cooling water from Mulben Burn.

PLANT: Semi-Lauter mash tun (11.5 tonnes). Eight stainless steel washbacks. Four lamp-glass wash stills (12,700 litres charge); four lamp-glass spirit stills (7,900 litres charge). Indirect fired. Shell-and-tube condensers.

MATURATION: Mainly refill hogsheads matured on site, where there is warehousing for 265,000 casks.

STYLE: Grassy.

MATURE CHARACTER: Sweet and lightly honeyed, with Sugar Puffs breakfast cereal. Cooked apples in the taste, with a whiff of smoke. Light- to medium-bodied.

Aultmore

REGION
Speyside
ADDRESS
Aultmore, by Keith,
Moray
PHONE
01542 881800
WEBSITE
www.aultmore.com
OWNER
John Dewar and Sons
(Bacardi)
VISITORS
By appointment
CAPACITY
1,200,000 L A

HISTORICAL NOTES: Alexander Edward of Forres, who had inherited Benrinnes Distillery from his father and led the consortium that built Craigellachie Distillery, built Aultmore in 1895, five miles north of Keith. The site, known as the Foggie Moss, was abundant in springs and rich in peat, and had been a favourite haunt of smugglers in the early nineteenth century. Aultmore went into production in 1897; so well-regarded was its make that capacity was doubled within a year (to 100,000 gallons per annum) and electric light installed, although all machinery in the plant was driven by either steam engine or water wheel. The wheel (which was used until about 1900) and an abortive 18hp steam engine are still there, although not in use.

Alexander Edward bought Oban Distillery in 1898 and floated The Oban and Aultmore-Glenlivet Distilleries Ltd, which was over-subscribed. Alas, one of the Directors was closely involved in Pattison's of Leith, whisky blenders, which went dramatically bankrupt in 1900. Production was cut, but Aultmore survived, to be put up for sale in 1923 and bought by John Dewar & Sons.

Dewar's joined D.C.L. in 1925, and from then until 1998 (when Dewar's and its associated distilleries was sold to Bacardi) it was managed by S.M.D. and its successors. They converted the stills to steam heating in 1967, closed Aultmore's maltings in 1968 and demolished and rebuilt the distillery in the 'Waterloo Street' style, adding two more stills to the existing pair in 1971. The make was not bottled by its owners as a single until 1996, although it has been ranked Top Class by blenders from the outset.

In 1998 Aultmore (and four other distilleries) went to Bacardi when that company bought John Dewar & Sons. Three new expressions were released in 2014 at 12, 21 and 25 years old, joined by an 18-year-old in 2015.

The illicit make from the Foggie Moss was popular among innkeepers in nearby Keith, Fochabers and Portgordon. One of the leading suppliers was Jane Milne. It was not uncommon for women to be illicit distillers in these days.

CURIOSITIES: The name means 'the big river' in Gaelic.

The power supply was generated by a water wheel from 1890 until 1960, and the wheel is still in situ – all that remains of the original buildings, which are described as 'a very modern, shining white complex' in the *Malt Whisky Yearbook*.

RAW MATERIALS: Process water from Auchinderran Burn; cooling water from the Burn of Ryeriggs. Unpeated malt from Burghead.

PLANT: Stainless steel Steineker mash tun (ten tonnes). Six larch washbacks. Two plain wash stills (charge 16,400 litres each) and two lamp-glass spirit stills (charge 15,000 litres each). All indirect fired. Shell-and-tube condensers.

MATURATION: 90% of the wood is American and 10% European, with more use of sherry-wood in old stocks. All the warehouses on site were demolished in 1996, the spirit now being tankered for maturation in Glasgow.

STYLE: Light and estery/fruity.

MATURE CHARACTER: Light and fragrant, with cut grass and green apples. The taste is sweet and floral – simple and easy. Light-bodied.

Balblair

REGION
Highland (North)
ADDRESS
Edderton, by Tain,
Ross-shire
PHONE
01862 821273
WEBSITE
www.balblair.com
OWNER
Inver House Distillers
VISITORS
By appointment
CAPACITY
1.8m L.P.A.

HISTORICAL NOTES: Balblair is among the oldest distilleries in Scotland, and one of the prettiest. Founded in 1790, by John Ross. Like many malts, the make was unknown beyond local farmers until relatively recently, but it is now – at last, for it deserves it – being nurtured and promoted as a brand by its owners.

The present distillery stands some distance from the original, six miles outside Tain, with its back to the Dornoch Firth. It was built around 1872, and expanded in 1894 (although it remains a small unit) to take advantage of the newly built railway line between Inverness and Wick

Balblair was managed by members of the Ross family for just over a century. They had also leased Brora and Pollo Distilleries from the Duke of Sutherland, and it was to the latter that Andrew Ross moved in 1896 (the distillery stood south of Tain, and closed in 1903).

Balblair reverted to the estate. Alexander Cowan became tenant and was obliged to rebuild. He went bust in 1911, and the distillery closed until 1947 when it was bought by a Banff solicitor, Robert Cumming (known as 'Bertie', he also owned Pulteney Distillery). He expanded it in 1964, building more warehouses, and changed from direct to indirect fired stills. In 1970 he sold to Hiram Walker-Gooderham & Worts of Canada and retired. Balblair accordingly came under the ownership of Allied (in 1988). They sold to the current owners, Inver House, in 1996.

Inver House was bought by a subsidiary of Thai Beverages plc (ThaiBev plc), Southeast Asia's largest alcoholic beverage company, in 2001 and integrated into that company's international division, International Beverage Holdings Ltd (InterBev) in 2006.

CURIOSITIES: Edderton is reputed to have the cleanest air in Scotland; hence the brand name *Elements*.

The new packaging makes good use of imagery related to the nearby Pictish stone known as the Clach Biorach ('the sharp stone'), on the Eadar Dun (hence Edderton). It was the 'Gathering Place' for the local community – a term that has been adopted by the 'Friends of Balblair', the malt's fan club.

Founded and managed for many years by Rosses, currently four out of the distillery's nine employees bear this surname.

The usually taciturn Charles Craig (in his *Scotch Whisky Industry Record*) described Balblair as 'one of the most attractive small distilleries still standing'.

In March 2007 Balblair packaging was elegantly redesigned by Curious Group, Glasgow.

Balblair featured in Ken Loach's award-winning and iconic film *The Angels' Share* (2012), as the scene of the auction of a cask of Malt Mill malt whisky (see *Lagavulin*).

RAW MATERIALS: Process and cooling water from the Allt Dearg Burn in the Struie Hills, piped five miles from the distillery. Floor maltings until the early 1970s, now unpeated malt from Portgordon Maltings.

PLANT: Semi-Lauter mash tun (4.6 tonnes). Six Douglas fir washbacks. One plain and dumpy wash still (19,600 litres charge) and one plain spirit still (11,800 litres charge), both indirect fired by steam. A heat converter system to pre-heat the wash was installed in 2006. The stillhouse also has a third small still, used as a model when the other stills were enlarged and converted to indirect firing. Shell-and-tube condensers.

MATURATION: Eight dunnage warehouses on site (one with a concrete floor installed by Norwegian troops during World War II, so it could be used as a canteen), with capacity for 28,500 casks. Mainly U.S. ex-bourbon hogsheads, some European oak. All single malt matured on site.

Edderton was known as 'The Parish of Peats', such was the ready supply of that fuel.

STYLE: Full-bodied, waxy-oily, nutty, leathery, but also green apples, floral notes and a natural spiciness.

MATURE CHARACTER: Nutty and sweet, with a trace of smoke and an elusive maritime note. The taste is medium-sweet, with fruity, nutty and spicy notes and an attractive thick mouthfeel. Medium- to full-bodied.

Ballindalloch

REGION
Speyside

ADDRESS
Ballindalloch Estate,
Banffshire

PHONE
01807 500331

WEBSITE
www.ballindallochdistillery.
com

OWNER
Ballindalloch Distillery LLP

VISITORS
By appointment

CAPACITY
100,000 L.P.A.

HISTORICAL NOTES: Ballindalloch Castle has been the home of the Macpherson-Grant family since 1546. Cragganmore Distillery stands on the estate, and was part-owned by the family between 1923 and 1965.

With the help of a Scottish government grant of £1.2 million, the family has installed a small, traditional distillery within a painstakingly restored farm steading, close to their recently created golf course. The plant was made by Forsyth of Rothes, including worm tubs, and all the restoration and conversion work was done by local tradesmen. Veteran distiller, Charlie Smith (formerly Glenkinchie, Talisker, etc.) was the consultant and Brian Robinson (ex-visitor centre manager at Glenfiddich and Glenlivet) looks after visitors. Production commenced in September 2014.

CURIOSITIES: The inspiration for the distillery came from a conversation after a game of golf over the adjacent Ballindalloch golf course, between Oliver Russell, the owner of the estate, Richard Forsyth, managing director of the famous coppersmiths in Rothes, and Douglas Cruikshank, production director at Chivas Brothers.

This is not the first licensed distillery on Ballindalloch estate. In the same year that the 1823 Excise Act – which laid the foundations of the modern whisky industry – became law, 'Delnashaugh Distillery' was licensed, the only one on Speyside. Next year it was followed by Balmenach, Cardow (Cardhu), Drumin (The Glenlivet), Macallan, Miltonduff and Mortlach.

The spirit safe, dating from 1863, was gifted to Ballindalloch by Cragganmore Distillery, which is also on the estate, and co-founded by Sir George Macpherson-Grant (see *Cragganmore*).

The visitor facilities comprise a 'long gallery' and a 'club room', and are conceived as an extension of the family home, Ballindalloch Castle, with portraits, bespoke furniture and a wood burning stove. Barley is grown on the estate, and the spirit matured here, allowing the Macpherson-Grants to describe Ballindalloch as the only 'single estate' distillery. (but see *Arbikie*).

RAW MATERIALS: Local barley from Ballindalloch estate, malted by Bairds, Inverness. Unpeated. Soft water from seven springs in the Garline Woods above the distillery.

PLANT: Stainless steel semi-Lauter mash tun (1 tonne). Four larch washbacks. One lamp-glass wash still (5,000 litres charge) and one boil-ball spirit still (3,600 litres charge), indirect fired. Worm tubs. External water wheel to help cool the cooling water.

MATURATION: Mainly ex-bourbon casks (ASBs and hogsheads; first-fill and refill), a small number of ex-sherry casks. Matured on site and at Marypark Farm, nearby.

STYLE: Full-bodied Speyside.

Balmenach

REGION
Speyside

ADDRESS
Cromdale, Moray

PHONE
01479 872569

WEBSITE
www.inverhouse.com

OWNER
Inver House Distillers

VISITORS
By appointment

CAPACITY
2.9m L.P.A.

HISTORICAL NOTES: One of the earliest modern authorities on Scotch whisky is Sir Robert Bruce Lockhart (author of *Scotch*, 1951). His great-grandfather, James MacGregor, farmed at Balmenach and distilled illegally there until, in 1823, he received a warning from the local Excise officer and took out one of the first licences on Speyside. James died in 1878, and was succeeded by his brother John (who came back from a successful career in New Zealand to take over).

When Alfred Barnard visited in the mid 1880s he described Balmenach as 'of the most antiquated type . . . [with] picturesque old pot stills and vessels'. This was by design: John MacGregor refused to allow any changes for fear that the character and quality of his whisky might be altered.

When his son, another James, took over in 1897, he formed a limited company, Balmenach-Glenlivet Ltd, and made some improvements, including a branch railway line with a steam locomotive that operated until October 1968, a few days before the Speyside line itself was closed. In 1922 the MacGregors sold to a consortium of blenders, which became part of D.C.L. in 1925.

Under S.M.D.'s management, Balmenach was extended from four to six stills in 1960 (an indication of the quality of the make, which is ranked First Class by blenders). In 1964 the floor maltings were converted to Saladin boxes. The mashhouse was rebuilt in 1968 and a dark-grains plant installed ten years later. All this was to the standard 'Waterloo Street' plan, with the difference that Balmenach was conceived so as to run exclusively by gravity, avoiding the need for pumps at any stage between charging the wash still and filling the casks (see *Caol Ila*).

Balmenach was mothballed in 1993 and sold to Inver House Distillers in 1997. Production was resumed the following year. The new owner

In 1879, the same storm that collapsed the Tay Bridge blew down the distillery chimney and the whole place nearly went up in smoke.

acquired no stock, which is why this excellent malt is not (yet) more readily available.

Inver House was bought by a subsidiary of Thai Bev plc in 2001, and integrated into that company's international division, International Beverage Holdings Ltd (InterBev) in 2006.

CURIOSITIES: A battle was fought near the distillery on the Haughs ('heights') of Cromdale in 1690, where a Jacobite force was surprised at night by a detachment of government dragoons and routed.

While it was closed during World War II (as most malt distilleries were), Balmenach was used as a billet by the Royal Corps of Signals.

For a period in the 1990s, Balmenach was bottled under the label Deerstalker by the Glasgow firm of Aberfoyle & Knight.

RAW MATERIALS: Soft water from a private supply, via the Cromdale Burn. Saladin boxes were installed in 1964 and used until the 1980s; thereafter unpeated malt from independent maltsters.

PLANT: Large cast iron semi-Lauter mash tun with copper dome (8.25 tonnes). Six Douglas fir washbacks. Three boil-ball wash stills (10,000 litres charge), three boil-ball spirit stills (10,000 litres charge). All indirect fired by steam. Worm tubs, each with 85-metre long worms. Dunnage warehouses hold 8,000 casks in total.

MATURATION: Mainly U.S. refill hogs; some Spanish oak.

STYLE: Meaty, vegetal, oily, full-bodied, some say slightly sulphury.

MATURE CHARACTER: The rich style of Balmenach makes it appropriate for European oak maturation, and the best bottlings have a dried fruit, sherried style, with a whiff of smoke. Medium-sweet, rich and dry in the finish. Full-bodied.

Balvenie

REGION
Speyside
ADDRESS
Dufftown, Moray
PHONE
01340 820373
WEBSITE
www.thebalvenie.com
OWNER
William Grant & Sons
VISITORS
V.I.P. tours available (must
be booked in advance).
Bottle Your Own facility,
shared with Glenfiddich.
CAPACITY
6.8m L.P.A.

HISTORICAL NOTES: William Grant opened Glenfiddich Distillery in 1886, and, during the next decade, distillery building on Speyside was unprecedented. Developers were soon trying to buy or rent land adjacent to Glenfiddich itself, so the Grant family pre-empted them by buying 12 acres of adjoining land, including New Balvenie Castle and Mains of Balvenie Farm. The Old Castle of Balvenie is a massive medieval keep that stands close by, its walls veiled by trees.

The new distillery, called Glen Gordon during its early months, went into production on 1 May 1893, with the old mansion house, which had lain empty and derelict for over 80 years, being converted into a maltings. In 1929 the upper storeys were demolished, and the lower storey turned into Warehouse 24. In 1956 the whole Balvenie Estate was bought by the Grant family.

The following year the distillery was enlarged to four stills; a further two were added in 1965; two more in 1971; one in 1991; and three in 2008.

CURIOSITIES: Balvenie House was actually a neoclassical mansion, commissioned in 1724 by William Duff, later 1st Earl of Fife, from the prominent architect James Gibbs, who had just completed the Church of St Martin in the Fields in Trafalgar Square. It is said that Duff built the house for 'a beautiful countess' (not his wife), who had been gifted a greyhound by a local admirer. The dog turned out to be rabid. It bit her and she soon succumbed to hydrophobia, as a result of which the house was abandoned by its owner. It had been lived in for only eight years, and had been derelict for 80 years when William Grant bought it.

In 1895 the building was owned by his great-nephew, Alexander, who was created first Duke of Fife in 1900, upon his marriage to Princess Louise, a daughter of Queen Victoria.

'Old' Balvenie Castle is often missed by visitors to the Genfiddich/Balvenie complex, but is well worth a visit. Its origins are shrouded in the mists of the Pictish twilight. There may have been a fortification here by 1010, when Malcolm Canmore defeated a Danish army at the Battle of Mortlach, and it was certainly used as a base by King Edward I of England ('The Hammer of the Scots') in 1296, when he was subduing the province of Moray. By the mid 1300s the castle was controlled by Alexander Stewart, Earl of Buchan and son of King Robert II, described as a 'ruffianly paranoiac' and known as 'The Wolf of Badenoch' on account of his rapacious behaviour – not least the burning of Elgin Cathedral, following a tiff with the Bishop of Moray.

Balvenie prides itself on 'doing everything in-house'. The distillery grows its own barley (albeit only a tiny amount) and has its own floor maltings (producing about 10% of its requirement), coppersmiths and bottling facility. Balvenie was first bottled as a single in the 1920s, then in the famous triangular bottle of the early 1970s (see *Glenfiddich*), followed by Founder's Reserve in a striking vintage champagne-style bottle in 1982. The same year, Balvenie Classic was released, in an equally stylish and unusual bottle. This makes a good claim to be the first 'double matured' malt, although the label does not say so.

In 2007 Grant's introduced a 'Bottle Your Own Balvenie' scheme, direct from the cask in Warehouse 24. There are three different casks to choose from – a sherry butt, a refill cask and a bourbon cask, all distilled in 1994. You draw the whisky with a 'dog', a copper tube sealed at one end with a coin and attached to a length of string. Such home-made devices were used by creative workers to steal whisky in days gone by. It was called a 'dog' on account of it being 'man's best friend', according to Dennis McBain, Balvenie's coppersmith! 'It was

usually hidden down the trouser leg, tied to the belt, so that it would not slip down and be revealed (this would result in immediate dismissal); he tells me. Sales of Balvenie have increased by 85% over the past decade, to nearly three million bottles, placing it as the eighth most popular single malt in the world.

RAW MATERIALS: Soft water from springs in the Conval Hills, some barley from Mains of Balvenie and the rest from independent maltsters. Peat from Tomintoul.

PLANT: Semi-Lauter mash tun (11.8 tonnes). Nine Oregon pine washbacks. Five boil-ball wash stills (three charged with 12,729 litres each and two with 9,092 each); four boil-ball spirit stills (charged with 12,729 litres each), all indirect fired. Shell-and-tube condensers.

MATURATION: Forty-four warehouses on site. American and European oak, with some port casks. For the 10-year-old, sherry-wood comprises 10% and ex-bourbon 90%.

STYLE: Fruity, full-bodied and honeyed.

MATURE CHARACTER: Rich and complex on the nose, with honeycomb, dried fruits (including orange peel) and some malt. Big mouthfeel and a sweet taste, drying, with light acidity. Medium- to full-bodied.

Banff Demolished

REGION
Highland (East)
ADDRESS
Inverboyndie by Banff,
Moray
LAST OWNER
D.C.L./S.M.D.
CLOSED
1983

*Banff Distillery
was always known
as Inverboyndie
locally*

HISTORICAL NOTES: The elegant Royal Burgh of Banff was granted its charter by Robert the Bruce in 1324. During the seventeenth and eighteenth centuries many of the local landed gentry had town houses there, and the town retains an atmosphere of faded gentility.

The first distillery here was founded in 1824 at Mill of Banff, but closed in 1863. Its then owner moved to a site at Inverboyndie, a mile west of the town, overlooking Boyndie Bay, to take advantage of the recently opened railway line. The original buildings were largely destroyed by fire in 1877, but were rebuilt 'within a wonderfully short time' (six months).

The owner, James Simpson & Company Ltd, went into voluntary liquidation during the Depression of 1932, and the distillery was bought by S.M.D., who closed it temporarily.

Some improvements were made during the 1960s, although the exterior remained unaltered. S.M.D. closed Banff Distillery in May 1983.

CURIOSITIES: During World War II the distillery was closed and the buildings used to billet soldiers from the King's Own Scottish Borderers. On the afternoon of 16 August 1941, a single enemy aircraft bombed the site. No one was injured, but a warehouse was burned to the ground; exploding whisky casks were seen to fly in the air, and those remaining were smashed to prevent the fire spreading. The *Banffshire Journal* reported that 'so overpowering were the results that even the farm animals grazing in the neighbourhood became visibly intoxicated'.

It was said that ducks dabbling in the Boyndie Burn were recovered at the seashore, some dead, some drunk, and that cows could not stand to be milked.

As well as the fire in 1877 and the attack in 1941, Banff Distillery suffered damage in 1959 when a still exploded. After decommissioning it was demolished section by section, the last part being cleared in 1991, following a further fire.

Ben Nevis

REGION
Highland (West)
ADDRESS
Lochy Bridge, Fort William
PHONE
01397 702476
WEBSITE
www.bennevisdistillery.com
OWNER
Ben Nevis Distillery Ltd (Nikka, Asahi Breweries)
VISITORS
Visitor centre, museum, exhibition, café
CAPACITY
2m L.P.A.

HISTORICAL NOTES: 'Long John' Macdonald, the founder of the distillery, was a big man and proud, and perhaps on account of his size and strength was chosen by the Lochaber lairds to establish a legal distillery near Fort William in 1825, when he was 27 years old. The original distillery produced only 200 gallons a week, but the make's reputation stretched even to Buckingham Palace, which accepted a cask in 1848, to be broached on the Prince of Wales' 21st birthday in 1863. After the death of 'the Old Gael' the business passed to his son, Donald Peter Macdonald. Although not as flamboyant a character as his father, it was Peter who really laid the foundations of the distillery's success.

He expanded and refurbished – several of the present-day buildings date from this time – and by 1864 was producing ten and a half times the amount of whisky than did his father. By 1877 he was employing 51 men and was marketing his product as Long John's Dew of Ben Nevis Pure Highland Malt Whisky. The following year he designed and built another larger 'model distillery' a mile away, at the mouth of the River Nevis, after which it is named, and ran it in tandem with its sister distillery.

Nevis Distillery had 'the largest Maltings under one roof in the North of Scotland' (the malting floor was 'so capacious that 3,000 persons could be seated there with ease') and also a substantial joiner's shop, a 'whole street of Workmen's Cottages' and a 'Warehouse and Stores' connected to the distillery's own harbour and wharf. All told, the distillery employed 200 men.

On Peter Macdonald's death in 1891, the enterprise passed to his sons but the 'whisky boom' would shortly turn to bust. Production at Nevis Distillery ceased in 1908 (and never resumed), although the warehouses were used until they were demolished to make way for sheltered housing in the mid 1990s.

'Long John'
Macdonald is
reputed to have
saved his brother
from being gored
by a bull by
seizing its horns,
wrestling it to
the ground and
dislocating its
neck.

Ben Nevis Distillery continued to operate off and on, and was then sold to the Canadian entrepreneur, Joseph Hobbs, in 1941. He paid £20,000 for the business, and the same day sold the warehousing at Nevis Distillery to Train & McIntyre (see below) for the same price! He only resumed production in 1955, having installed a continuous still alongside his pot stills, concrete washbacks and a novel malt handling system.

Production continued until 1978. In 1981 Hobbs' son sold it to Long John International, the spirits division of Whitbread plc, who embarked on a programme of modernisation, completed in 1984. The general downturn in the whisky trade at the end of the 1980s obliged Long John to sell in 1989. (Whitbread itself was taken over by Allied-Lyons the same year, to form Allied Distillers.) The new owner was Nikka Whisky Distilling Company of Japan, a company founded by Masataka Taketsuru, 'the Father of Japanese Distilling', who had trained in Scotland shortly after World War I.

CURIOSITIES: 'Long John' Macdonald was a man who attracted tales. In one account, he routed a band of smugglers who ambushed him, resentful of his having 'gone legal', armed with nothing but a stout *cromach* (crook).

During the year 1884–85, when the two Nevis distilleries produced 260,000 gallons of whisky, Macallan Distillery produced around 40,000, Glen Grant 140,000 and The Glenlivet just under 200,000; in 1889 a newspaper article describes the Ben Nevis Distilleries as being 'almost double the size of their nearest rival'.

Joseph Hobbs made a fortune running Scotch into the United States from Canada during Prohibition. One of his main suppliers was Teacher's, for whom he shipped 137,927 cases of Highland Cream via Antwerp into San Francisco Bay

on HMCS *Stadacona*, a converted Canadian naval vessel. Another of his steam ships, *Ocean Mist*, is now a rather drab floating night-club, permanently moored at The Shore in Leith. (See also *Glenlochy*.)

Soon after World War II, Joe Hobbs bought Inverlochy Castle – initially with the intention of turning it into a distillery (it had been billeted and was in a poor state of repair) from its owner, the 6th Lord Abinger. His lordship's wife was Marguerite Steinheil, who had been the mistress of the French president Félix Faure, who died in her arms in 1899. In 1908, she caused an even more dramatic scandal when she was found (insecurely) tied to a bed, with the bodies of her husband and stepmother, the former strangled, the other choked by her false teeth. She was acquitted of murder – in the words of John Julius Norwich (in *The Times*, January 2014) she was: 'Far too beautiful to be found guilty . . . and in 1917 married Lord Abinger, leading thereafter a life of incorruptible rectitude.' Inverlochy Castle is now one of Scotland's very best hotels.

Joe Hobbs pioneered the process of 'blending at birth', that is, blending malt and grain new-make spirits and maturing it together. When asked what malts he used, he is said to have replied, 'Oh, just whatever's around'! By the 1950s Dew of Ben Nevis had become a cheap blend.

Fascinating though the story of Ben Nevis Distillery is, the site itself has been sadly neglected. *Whisky Magazine* described it as reminiscent of 'a fading Siberian tractor collective'!

RAW MATERIALS: Water from Allt a'Mhuilinn (Mill Burn), which flows from Coire Leis and Coire na'Ciste near the top of Ben Nevis itself. Unpeated malt from independent maltsters. Brewers' yeast.

PLANT: Stainless steel full-Lauter tun (8.5 tonnes). Six stainless steel washbacks, two Oregon pine. Two plain wash stills (21,000 litres charge), two plain spirit stills

(12,500 litres charge). Indirect fired. Shell-and-tube condensers.

MATURATION: A combination of ex-bourbon and ex-sherry casks, on site in five traditional dunnage and one racked warehouse.

STYLE: Malty and robust, with a hint of smoke.

MATURE CHARACTER: Aromatic, fruity (cooked fruits), malty, with dark chocolate notes. Mouthfeel creamy; sweet start with vanilla and caramel, a trace of sulphury smoke. Medium- to full-bodied.

BenRiach

REGION
Speyside
ADDRESS
Longmorn, by Elgin,
Moray
PHONE
01343 862888
WEBSITE
www.benriachdistillery.
co.uk
OWNER
The BenRiach Distillery
Company Ltd
VISITORS
By appointment
CAPACITY
2.8m L.P.A.

BenRiach was named Distillery of the Year in the 2007 Malt Advocate Whisky Awards.

HISTORICAL NOTES: Until 2004, BenRiach's fortunes were closely linked to its larger and more famous neighbour, Longmorn. It was built by the owner of that distillery, John Duff, in 1897, and designed by Charles Doig, the leading distillery architect of the day. Duff ran into financial trouble and sold it to Longmorn Distillers Company two years later. Mothballed in 1900 (although it continued to supply malt to its neighbour, to which it was connected by a quarter-mile-long railway line), it was only opened again in 1965, now owned by The Glenlivet Distillers Ltd, who had rebuilt it. Seagram's acquired The Glenlivet Group in 1977, and the licence was transferred to its subsidiary Chivas Brothers.

Seagram's whisky division was bought by Pernod Ricard in 2001, and BenRiach was again mothballed until 2004, when it was sold to a consortium of three entrepreneurs, led by veteran distiller Billy Walker, with backing from the South African company, Infra Trading, for £8 million. It went back into production in September that year.

In April 2016 it was announced that BenRiach, along with its sister distilleries Glendronach and Glenglassaugh (see entries) had been bought by the large American distiller Brown-Forman (owner of Jack Daniels) for £285 million.

CURIOSITIES: Billy Walker is one of the most experienced people in the trade today. Trained as an organic chemist, he joined Ballantine's in 1971. He then went to Inver House as a blender, then on to Burn Stewart Distillers. He was a member of the management buy-out team that acquired Burn Stewart in 1988 (see *Deanston*), becoming its Production Director. In July 2008 he bought Glendronach Distillery from Chivas Brothers (see *Glendronach*), and in 2013 bought Glenglassaugh Distillery from the Scaent Group.

Although mothballed for much of its existence, BenRiach's floor maltings were in almost continuous use until 1999, with peat rights on Faemussach Moor, near Tomintoul. Even during the Seagram ownership, malt for Longmorn was made here, while BenRiach itself relied on malt from outside sources. After 1980, when the railway lines between the two distilleries were lifted, BenRiach used its own malt. The new owners re-opened the maltings in 2013, having done trial runs since November 2012.

Before 1965 BenRiach had one spirit still and one wash still. This was then doubled, and a third spirit still installed at a later date. This created an imbalance, however, so it was taken out (and ultimately sold to a Canadian business).

From 1983, BenRiach produced a quantity of peated malt as well as its usual unpeated style each year – a unique move at the time for a Speyside distillery, but one that allows the new owners to have aged peated Speysides in their portfolio.

RAW MATERIALS: Process water from deep springs about half a mile away, known as Burnside, shared with Longmorn; cooling water from the Glen Burn (which also supplies Glen Elgin, Longmorn and Linkwood Distilleries). Currently peated (*c.* 35ppm phenols) and unpeated malt from independent maltsters.

PLANT: Traditional stainless steel mash tun with raised canopy (5.8 tonnes). Eight stainless steel washbacks. Two plain wash stills (15,000 litres charge), two plain spirit stills (9,600 litres charge). Indirect fired. Shell-and-tube condensers.

MATURATION: Now all filled into fresh ex-bourbon barrels for five to six years, then re-racked into refill hogsheads and butts.

STYLE: Sweet and fruity, Speyside style; sweet and smoky.

MATURE CHARACTER: (for the unpeated, unfinished malts) Fruity and estery, with apples and green bananas, some cereal notes. Sweet and creamy to taste, with vanilla and light caramel. Light- to medium-bodied.

Benrinnes

REGION
Speyside
ADDRESS
Aberlour, Moray
PHONE
01340 872600
WEBSITE
www.malts.com
OWNER
Diageo plc
VISITORS
By appointment
CAPACITY
3.5m L.P.A.

Benrinnes is ranked Top Class by blenders, and is extensively used in Diageo's blends.

HISTORICAL NOTES: A distillery was established 700 feet above sea level on the northern slopes of Ben Rinnes, at Whitehouse Farm, by one Peter McKenzie in 1826. It was destroyed by the Great Flood of Moray two years later and another distillery was built at Lyne of Ruthrie some distance away by John Innes.

Innes went bankrupt in 1834. His successor, William Smith, also failed 30 years later (and was sent to prison), and the lease passed to David Edward, farmer, and then to his son Alexander, who became a well-known distiller, founder of Aultmore, Dallas Dhu and Craigellachie Distilleries, part-owner of Oban and Yoker and supporter of Benromach, which was built on his estate near Forres.

The distillery suffered what the *Northern Scot* described as a 'rather destructive fire' in 1896, but its fortunes were more severely hit by the failure of its agents, F.W. & O. Brickmann, in the fall-out of the Pattison crash of 1899, followed by the general recession in the whisky industry.

Alexander Edward sold to Dewar's in 1922, and the distillery passed to D.C.L. in 1925, with management by S.M.D. from 1930. It was rebuilt in 1955/56, and doubled in size to six stills ten years later. The 'new' half runs independently of the older half, and the two are vatted prior to filling into cask. Saladin boxes replaced floor malting in 1964, and operated for 20 years; steam heating replaced direct firing in 1970.

CURIOSITIES: Since 1956, Benrinnes has practised an unusual form of 'partial triple distillation', its six stills being arranged in two sets of three, so that some of the spirit is distilled three times and some of it twice.

Each wash still is charged with the contents of one washback. This distillation is split into heads and tails, the stronger heads being forwarded to the

Strong Low Wines Receiver, and the weaker tails to the Weak Low Wines Receiver. The intermediate spirit still is charged from the Weak Low Wines Receiver and again the distillation is split: the heads going to the Strong Low Wines Receiver and the tails to the Weak Low Wines Receiver. The spirit still is charged from the Strong Low Wines Receiver, with the foreshots being transferred back to the same receiver, and the aftershots/feints being further split: strong feints going to the Strong Low Wines Receiver and weaker feints to the Weak Low Wines Receiver.

It might be expected that this would produce a light spirit, but the reverse is true, the effect of triple distillation being countered by the size and shape of the stills and the use of worm tubs.

RAW MATERIALS: Floor maltings until 1964, then Saladin boxes until 1984. Unpeated malt now comes from Burghead and Roseisle. Process water from the Scurran and Rowantree Burns, which rise high up on Ben Rinnes itself.

PLANT: Three-arm semi-Lauter mash tun (8.7 tonnes). Eight larch washbacks. Two plain wash stills (20,000 litres charge), two plain intermediary stills (5,243 litres charge), two plain spirit stills (7,000 litres charge). All indirect fired by steam. Worm tubs.

MATURATION: Mainly European oak and matured off site.

STYLE: Heavy, meaty.

MATURE CHARACTER: Benrinnes has a famously big, robust character. The nose is rich with burnt caramel, dried fruits, sherry notes. The texture in the mouth is big, filling, velvety. The taste is sweet, then drying, with a long finish. Full-bodied.

Benromach

REGION
Speyside

ADDRESS
Forres, Moray

PHONE
01309 675968

WEBSITE
www.benromach.com

OWNER
Gordon & MacPhail

VISITORS
Malt Whisky Centre
opened 1999
(4* attraction awarded
by the Scottish Tourist
Board). Fill your own
bottle facility

CAPACITY
500,000 L.P.A.

HISTORICAL NOTES: Benromach has had a chequered career, but is hopefully now on a sound footing.

The Benromach Distillery Company was founded in 1898 by Duncan MacCallum (of Glen Nevis Distillery, Campbeltown) and F.W. Brickmann (spirit dealer, Leith), with the support of Alexander Edward, who rented them the site on the northern edge of his Sanquhar Estate, Forres. (See *Aultmore*.)

As the distillery was nearing completion, Pattison's of Leith, one of the largest buyers of new-make spirit, suspended payments. Brickmann's firm was closely associated with Pattison's and ceased trading, with liabilities of over £70,000 (£7.7 million today). Benromach did not go into production until 1909.

It was sold in 1911 to a London company, closed from 1914 to 1919, then acquired by Alloa brewer John Joseph Calder, who immediately sold it to Benromach Distillery Ltd (a company owned by Macdonald, Greenlees & Williams, of Leith, and six English breweries).

It worked for a spell in the mid 1920s, but 'had been silent for years' by 1937, when it was bought

by Associated Scottish Distillers (a company formed by the colourful Joseph Hobbs (see *Ben Nevis*, *Glenesk* etc.), and sold next year to National Distillers of America. They sold to D.C.L. in 1953, and Benromach was transferred to S.M.D.

Benromach was refurbished in 1966, when the stills were converted to indirect firing; the floor maltings were closed in 1968. It was mothballed in 1983, and the plant removed. Ten years later it was sold to Gordon & MacPhail (see 'Leading Independent Bottlers', in *Facts and Figures*), and carefully refurbished. All plant was replaced except a spirit receiver. The new distillery was opened by Prince Charles, Duke of Rothesay, on 14 October 1998; the visitor centre opened the following year.

CURIOSITIES: Benromach was formerly a much larger affair, but by the time Gordon & MacPhail bought it, nothing remained of the former plant except one receiver tank. During the five years between purchase and opening, the new owners did many experimental distillations before settling on the style they wanted. After they had gone into production, the former owner, Diageo, offered them a box of new-make samples from pre-1993. The spirit character was almost identical, and yet the distillery had been rebuilt. The only constants were the location and the water ...

In 1925 it had a rare (perhaps unique) wooden mash tun.

When S.M.D. took over, it was remarked that Benromach had a particularly good external appearance, enhanced by gardens laid out by the buccaneering Joe Hobbs. A key feature today is its smart red brick chimney, contrasting nicely with the whitewashed walls of the distillery.

Since they acquired the distillery, Gordon & MacPhail have released a large number of expressions, at various ages and finishes.

RAW MATERIALS: Malt from independent maltsters in Scotland, peated to around 10ppm phenols, with some experimental distillations at heavier and lighter peating. Process water from Chapelton Spring; cooling water from the Burn of Mosset, which rises on Romach Hill. Benromach is unusual, possibly unique in Scotland, in continuing to use both brewers' yeast and distillers' yeast.

PLANT: Semi-Lauter mash tun (1.5 tonnes); four Scottish larch washbacks. One plain wash

'With its high-pitched gables and mullioned windows in the Scots vernacular style of the seventeenth century, [it] surprised and delighted the eye.'
Brian Spiller

still (7,500 litres); one boil-ball spirit still (5,000 litres). Both indirect fired; both shell-and-tube condensers. Spirit safe from Millburn Distillery. External shell-and-tube condensers.

MATURATION: Mainly refill U.S. hogsheads, some refill European wood. Matured on site and in Elgin.

STYLE: Fruity, with body.

MATURE CHARACTER: Light Speyside in style, with more body than some. Fresh and fruity/floral on the nose; creamy to taste; sweet overall with some sweet cereal notes. Light- to medium-bodied.

Ben Wyvis Dismantled

REGION
Highland (North)
ADDRESS
Invergordon, Ross &
Cromarty
LAST OWNER
Invergordon Distillers
CLOSED
1977

Only two bottles of the original Ben Wyvis (and one half-bottle), dating from the 1890s, have appeared at auction.

HISTORICAL NOTES: There were two Ben Wyvis distilleries. The first was built in 1879 on the outskirts of Dingwall – some of the buildings still stand, converted into flats. In 1893, its name was changed to Ferintosh, and it closed in 1927, although the warehouses were in use until 1980.

The second Ben Wyvis was situated within Invergordon grain whisky distillery, nearby. This malt distillery was installed in 1965, with two stills and a capacity of 750,000 L.P.A. It was silent from 1977 and has now been dismantled.

There have only been a handful of bottlings of Ben Wyvis (Invergordon), and all of these very limited.

CURIOSITIES: Ben Wyvis was the model for Tamnavulin Distillery, also built by Invergordon, in 1966.

Bladnoch

REGION
Lowland
ADDRESS
Bladnoch, Wigtown,
Wigtownshire
PHONE
01988 402605
WEBSITE
www.bladnoch.co.uk
VISITORS
Yes
CAPACITY
250,000 L.P.A.

In 1792 a schooner was arrested on the Solway Firth for smuggling contraband. Among the Excise officers involved was Robert Burns.

HISTORICAL NOTES: Bladnoch was established as a farm distillery by the brothers Thomas and John McClelland in 1817, and operated by John's son, Charlie, until 1905 when production ceased. It was sold to the Irish distillers Dunville & Company in 1911, and between that date and 1937 (when Dunville went into liquidation) it operated only intermittently. It was bought by Ross & Coulter of Glasgow, who dismantled it in 1941, sold the stock (89,000 gallons of whisky) at below its value – which led to the Inland Revenue imposing 100% 'excess profits tax' – and sold the plant to Sweden (one still is now in a museum). The distillery buildings went to A.B. Grant. Trading as Bladnoch Distillery Company, he installed two new stills in 1956 and sold to McGowan and Cameron, whisky blenders in Glasgow, ten years later, who doubled the distillery's capacity (to four stills).

Inver House owned the distillery for ten years until 1983, then sold it to Arthur Bell & Sons, so it became part of Guinness in 1985, and of U.D. in 1987. They reduced capacity in 1986 by disconnecting a pair of stills, then closed and decommissioned it altogether in June 1993. The local authority continued to manage some of the buildings as a 'heritage centre'.

This should have been the end of the road for Bladnoch. However, in 1994 it was bought by a developer from Banbridge, Northern Ireland, a family company directed by Raymond Armstrong, his brother Colin and their wives. The original plan was to develop the site as holiday cottages, but the new owners quickly realised what an important role the distillery had played in the local economy. Also, the distillery visitor centre was successful, yet without a distillery it would soon become pointless. U.D. had sold Bladnoch on the understanding that it would never be brought back into production, but they were prepared to waive this (to a degree) and help the new owner bring the distillery alive again. Bladnoch went back into production in December 2000, but ceased again in 2009.

The company went into voluntary liquidation in 2014 and in July 2015 was bought by an Australian entrepreneur, David Prior, who had recently sold his highly successful yoghurt company for $80 million. He has undertaken substantial refurbishment and plans to increase capacity to one million litres. Veteran master blender, Ian MacMillan, is the distillery manager.

CURIOSITIES: Wigtown is on the Solway Firth, at the southern extremity of Scotland, and Bladnoch is the most southerly of Scottish distilleries, well off the beaten track – although the district once boasted 11 distilleries.

Bladnoch Distillery plays a crucial and integral role in the local community, welcoming around 25,000 visitors a year.

RAW MATERIALS. Process and cooling water (soft and peaty) from the River Bladnoch. Mainly unpeated malt from independent maltsters. Very occasional medium-peated (18ppm phenols) malt, usually once a year.

PLANT: Semi-Lauter stainless steel mash tun (five tonnes). Six Oregon pine washbacks. One boil-ball wash still (13,500 litres charge), one boil-ball spirit still (10,000 litres charge) with sight glasses (it was once a wash still). Indirect fired by steam. Shell-and-tube condensers.

MATURATION: Thirteen dunnage-style warehouses on site. 80% first-fill bourbon barrels (4 Roses and Heaven Hill), sherry butts and hogsheads, and occasional ex-wine casks. Unusually high levels of evaporation.

STYLE: Sweet, grassy, malty, mellow.

MATURE CHARACTER: Pastoral and floral; reminiscent of hedgerows, with lemony/citric notes and some cereal. The taste echoes this. The overall impression is light and appetising. Short finish.

Blair Athol

REGION
Highland (South)

ADDRESS
Perth Road, Pitlochry

PHONE
01796 482003

WEBSITE
www.malts.com

OWNER
Diageo plc

VISITORS
Visitor centre built 1987

CAPACITY
2.8m L.P.A.

HISTORICAL NOTES: Adjacent to the distillery's rustic stone buildings, some thickly clad in Virginia creeper, flows the Allt Dour ('burn of the otter'), and this stream gave its name to the first distillery on the site, Aldour, founded in 1798. The name was changed to Blair Athol in 1825 when it was rebuilt and expanded.

In 1882 the distillery was bought and extended again by the Edinburgh blenders Peter Mackenzie & Company (who later founded Dufftown Distillery). It was mothballed between 1932 and 1949, then tastefully restored by Arthur Bell & Sons, who had bought P. Mackenzie & Company in 1933. Professor McDowell described it in the 1960s as 'almost a model distillery'.

Bell's first refurbishment in 1949 cost £75,000; in 1970 they doubled the distillery's capacity, expanding from two to four stills and took out the floor maltings. Bell's Extra Special blended Scotch was by now the best-selling brand in the U.K.; the output of Bell's three distilleries quintupled between 1960 and 1970.

But Arthur Bell & Sons fell victim to its own success. In 1985 it was subject to a hostile take-over by Guinness plc and soon after it became part of U.D. (now Diageo).

CURIOSITIES: After the Battle of Culloden a local laird, Robertson of Faskally, who had been a captain in the Jacobite army, returned to his own country with a price on his head. He hid in Allt Dour farmhouse, then escaped down the Allt Dour Burn and sheltered in an old oak tree near the distillery until the red coats had passed on. Then he escaped to France.

With the acquisition of Blair Athol in 1933, via Peter Mackenzie & Company, Arthur Bell & Sons moved from being a small local blender to being a medium-sized distiller with the potential to become a major player.

Blair Athol is the heart-malt for the Bell's blends.

RAW MATERIALS: Floor maltings until 1960s; now unpeated malt from Glen Ord. Water from the Allt Dour, which rises as a spring in the heights of Ben Vrackie ('the speckled hill'), above the snow line.

PLANT: Semi-Lauter mash tun (eight tonnes). Four stainless steel washbacks; four Oregon pine washbacks (these came from Mortlach in the 1980s). Two plain wash stills (13,000 litres charge), two plain spirit stills (11,500 litres charge). Shell-and-tube condensers, run hot, with after-coolers.

MATURATION: Mainly ex-bourbon refills, some ex-sherry casks used in the proprietary single malt bottlings. Mainly matured in the Central Belt.

STYLE: Rich, nutty-spicy and cereal-like.

MATURE CHARACTER: Nutty and caramel-like on the nose, with rich maltiness, traces of leather and tobacco. A rich whisky, it takes European oak maturation well, adding to richness, with fruit cake, traces of wine and some sulphury notes. A curious mix of sweet and dry to taste, and not as long in the finish as the nose suggests. Medium-bodied.

Bowmore

REGION
Islay

ADDRESS
School Street, Bowmore,
Isle of Islay, Argyll

PHONE
01496 810441

WEBSITE
www.bowmore.com

OWNER
Beam Suntory

VISITORS
New visitor centre
opened 2007, cottages
available for let

CAPACITY
2.2m L.P.A.

HISTORICAL NOTES: Bowmore is the oldest distillery on Islay, and one of the oldest in Scotland. The usual date ascribed to its foundation is 1779, but the distillery may have been established a decade before this, when the model village of Bowmore was built by Daniel Campbell of Shawfield. David Simpson was brought over from Bridgend to build the distillery and was succeeded by his relation Hector Simpson. Hector sold to James and William Mutter, Glasgow merchants of German extraction, in 1837. They expanded the distillery and their sons retained ownership until 1892.

Ownership changed hands several times between then and 1925 when the distillery was purchased by a Skye man, Duncan MacLeod, trading as J.B. Sherriff & Company. (They had owned the distillery briefly, before going into liquidation in 1920, and had also owned Lochhead Distillery in Campbeltown and Port Charlotte Distillery, Islay.) Sherriff's sold to William Grigor & Son of Inverness, who had rebuilt Glen Albyn Distillery there in 1884.

During World War II Bowmore was requisitioned by the Air Ministry to support Coastal Command's efforts to protect Atlantic convoys.

In 1963, the Glasgow whisky broker Stanley P. Morrison bought the distillery for £117,000 and began modernising and expanding the site, including an innovative waste-heat recovery system, which was estimated to save just over £100,000 per annum at the time it was installed in 1983. Hot water from the condensers pre-heated the wash, heated the malt kiln and warmed the water for the local public swimming pool, located within a former bonded warehouse, donated to the community by Morrison Bowmore in 1983.

Suntory bought a 35% stake in the company in 1989 and whole ownership in 1994.

CURIOSITIES: Bowmore is one of the few distilleries to retain its own floor maltings, producing around

Bowmore whiskies are highly collectable. A bottle engraved 'W. & J. Mutter 1890' achieved £13,000 in 2001; another 'Mutter' bottle, allegedly from 1851, was bought for £25,000 in 2007.

30% of its malt requirement. It was one of the first distilleries to install shell-and-tube condensers, in 1886.

The whisky made here has long had a high reputation. As early as 1841, Walter Frederick Campbell of Islay, the laird, received an order from Windsor Castle to supply 'a cask of your best Islay Mountain Dew' for the Royal Household – cask size and price of no concern, 'but the very best that can be had'. The order was renewed two years later. The Mutter brothers owned a steamship to carry casks up to Glasgow, where they were bonded beneath the arches of Central Station.

In September 2012, Bowmore announced the release of 12 bottles of a 54YO expression, distilled in 1957, to be auctioned for Scottish charities in October that year, in New York, with a reserve price of £100,000. At the time, this was the highest price achieved by a bottle of malt whisky.

RAW MATERIALS: Soft peaty water from River Laggan, seven miles from the distillery, for process and cooling. Three malting floors produce 30% of the malt requirement; the remainder comes from Simpson's of Berwick, peated to 25ppm phenols. 100% Scottish malt.

PLANT: Stainless steel semi-Lauter mash tun, with copper cover (eight tonnes). Old riveted 'coppers' to charge the mash tun. For a period, these were replaced with stainless steel, then reverted. Six Douglas fir washbacks. Stainless steel wash charger, to free up a washback (for five to six hours) prior to charging the wash still. Two plain wash stills (20,000 litres charge each). Two plain spirit stills, with sight glasses (14,637 litres charge each). All indirect fired. External shell-and-tube condensers run very hot to provide the heat recovery system referred to above. Internal after-coolers to complete condensation.

MATURATION: Mix of first-fill American barrels and hogsheads, first-fill sherry butts and puncheons. Each expression has its own mix of cask-types. A three-floor dunnage warehouse on site, incorporating the ancient Bowmore Vaults on its ground floor, lies partly below sea level, and two warehouses (one dunnage, one racked) are on the Low Road outside Bowmore village. 27,000 casks in total.

STYLE: Smoky and floral.

MATURE CHARACTER: In a blind tasting, my identifier for Bowmore is 'lavender'. The character varies a lot, depending on cask selection (which varies from expression to expression). Behind the lavender/air-freshener note, it is sweet, rich, fruity (a combination of exotic fruits like mango, passionfruit and dried fruits), malty and scented – smoky on the nose. A good texture; sweet taste with a belt of smoke (especially in the younger expressions), and still this lingering perfume. Medium-bodied.

Braeval

REGION
Speyside

ADDRESS
Chapeltown,
Ballindalloch, Moray

PHONE
01542 783042

WEBSITE
No

OWNER
Chivas Brothers

VISITORS
No

CAPACITY
4m L.P.A.

HISTORICAL NOTES: At one time, 36 distilleries attached the suffix 'Glenlivet' – for example, Macallan-Glenlivet, Aberlour-Glenlivet – although there are only three distilleries in the glen of that name: The Glenlivet itself, Tamnavulin and Braeval.

Braeval is the highest distillery in Scotland.

Originally named Braes of Glenlivet, it was built by Seagram's in 1973 to produce fillings for the Chivas blends. Like its sister, Allt-a-Bhainne, it is attractively laid out – uncompromisingly modern, but with traditional elements such as a pagoda roof. It is highly automated and can be run by one person. To start with, it had three stills, each with a distinctive bulge in the neck known as a 'Milton Ball'; two more were added in 1975 and a further one in 1978. There are no warehouses on site, all the make being tankered to Keith Bond for filling into casks, then to other sites for maturation; pot ale used to go to The Glenlivet Distillery to be evaporated into pot ale syrup, and draff is sold to merchants.

Pernod Ricard bought Seagram's in 2001 and Braeval was mothballed the following year. It re-opened in July 2008.

The whisky has never been bottled by its owner, only occasionally by independent bottlers.

CURIOSITIES: The district in which the distillery stands is known as Eskemulloch (which literally means 'headwater'). It is a mile north of the old Roman Catholic seminary of Scalan, one of the very few places in Scotland where priests could be trained after the Reformation. It was established by the (Catholic) owner of the glen, the Duke of Gordon, in 1717, but was destroyed by Hanoverian troops in 1746.

The Ladder Hills, from where Braeval draws its water, are crossed by the Whisky Road, used by smugglers in days gone by to carry their wares out of Glenlivet and down to the Lowlands.

Braeval is the highest distillery in Scotland.

RAW MATERIALS: Unpeated malt from independent maltsters. Water from the Preenie and Kate's Well Burns and the burn at Ladderfoot (once popular with smugglers).

PLANT: Traditional rake-and-plough mash tun with copper canopy (nine tonnes); 15 stainless steel washbacks. Two plain and stocky wash stills (22,000 litres each). Two boil-pot spirit stills (7,500 litres) – one run from the wash still charges two spirit stills. All indirect fired by steam. Shell-and-tube condensers.

MATURATION: Mainly refill U.S. hogsheads, all at Mulben, near Keith.

STYLE: Sweet and grassy.

MATURE CHARACTER: Most of the independent bottlings of Braeval have come from ex-sherry butts, which cover the distillery character. The nose is sweet, with fruit cake, chocolate and sherry; the taste rich and full, with spicy notes.

Bottlings from ex-bourbon refills, which retain the distillery character, are lighter and drier, with Speyside character and a sweet, fresh-fruity flavour. Medium-bodied.

Brora Dismantled

REGION
Highland (North)
ADDRESS
Brora, Sutherland
LAST OWNER
D.C.L./S.M.D.
CLOSED
1983

HISTORICAL NOTES: For most of its life Brora Distillery was called Clynelish, but in 1967 a new distillery was built next door and given the same name. The original was mothballed for a year, then resumed production as Clynelish No. 2 until 1975 when it was renamed Brora. It was closed in May 1983 and most of its plant removed. The elegant old buildings still stand and the warehouses are used by Clynelish.

Clynelish/Brora was founded in 1819 by the Marquess of Stafford (later first Duke of Sutherland) as part of his plan to 'improve' his wife's vast northern estates – a plan which also required the clearing of some 15,000 tenants from the land to make way for sheep. Some of those cleared were moved to new coastal townships like Brora.

For the first 70 years of its existence the distillery struggled. Several of its licensees filed for bankruptcy. Then, in 1896, it was acquired and rebuilt by James Ainslie & Company, blenders in Glasgow, in partnership with John Risk (owner of Bankier Distillery, near Falkirk); in 1912 the latter acquired complete ownership, in partnership with D.C.L. and (after 1916) with John Walker & Sons. Risk was bought out in 1925, when Walkers (and Clynelish/Brora) amalgamated with D.C.L.

Thereafter, the make was used in the Johnnie Walker blends. The distillery's two pot stills were converted from direct coal firing to internal steam heating in 1961, and the steam engine and water wheel were replaced by electric power in 1965, which is also the year when the floor maltings were used for the last time.

Between May 1979 and July 1983 Brora/Clynelish produced a heavily peated malt for blending purposes.

CURIOSITIES: The great Victorian connoisseur Professor George Saintsbury, author of *Notes from a Cellar Book*, wrote highly of Brora/Clynelish.

The distillery and its warehouses are listed

buildings and much of the original equipment –
stills, washbacks, mashtun – is still in situ.

Indeed, the trade journal *Harper's Weekly*
remarked in 1896 that 'the make has always
obtained the highest price of any single Scotch
whisky. It is sent out, duty paid, to private customers
all over the kingdom; and it also commands a very
valuable export trade; the demand for it in that way
is so great that the proprietors . . . have for many
years been obliged to refuse trade orders.'

In 2002, Diageo began to release limited annual
bottlings of Brora at 30YO (25YO after 2008), and in
January 2014, a 40YO Brora (travel retail exclusive;
260 bottles only; RRP £6,995) – the company's 'most
expensive malt whisky ever'.

Bruichladdich

REGION
Islay

ADDRESS
Bruichladdich, Isle of
Islay, Argyll

PHONE
01496 850221

WEBSITE
www.bruichladdich.com

OWNER
Rémy Cointreau

VISITORS
Visitor centre, café, shop
and whisky school

CAPACITY
1.5m L.P.A

HISTORICAL NOTES: Bruichladdich (pronounced Brewickladdie) has enjoyed a renaissance since 2000, when it was bought for £6.5 million by a private concern led by Mark Reynier, a wine merchant in London, with the support of local investors. During the previous decade the distillery had operated only part-time (with periods of complete closure) under the ownership of Whyte & Mackay (J.B.B. Greater Europe), who had acquired it when they took over Invergordon Distillers in 1993.

Bruichladdich was purpose-built from scratch in 1881 by the Harvey brothers (owners of Dundashill and Yoker distilleries in Glasgow) using the then new material, concrete, to bind beach pebbles. When William Harvey died in 1937, the distillery was sold to Joseph Hobbs, who ran it through his company Associated Scottish Distillers (A.S.D.) until it was sold in 1952 to Ross & Coulter, whisky brokers in Glasgow and owners of Bladnoch Distillery, for £205,000 (the value of the stock). R&C had previously sold Fettercairn Distillery to A.S.D. The firm was absorbed by D.C.L. in 1954 and wound up in 1960, when Bruichladdich was sold to A.B. Grant & Co, who had bought Bladnoch from R&C in 1956.

Grant's sold Bruichladdich to Invergordon Distillers in 1968 for £400,000. Capacity was doubled (to four stills) in 1975, then the distillery was mothballed during the grim years of the mid 1980s and again from 1995 to 1998, after Invergordon had been bought by Whyte & Mackay (1993).

Production in 1998/99 was minimal, and Bruichladdich only resumed full operation in July 2001, under its new owners. Over the next 11 years they worked tirelessly to build the brand, with hundreds of expressions. So successful were they that they were able to sell the distillery to the large French distillery Rémy Cointreau, in July 2012, for £58 million.

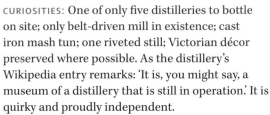

At 169ppm phenols, Octomore J. is the most heavily peated spirit in the world!

CURIOSITIES: One of only five distilleries to bottle on site; only belt-driven mill in existence; cast iron mash tun; one riveted still; Victorian décor preserved where possible. As the distillery's Wikipedia entry remarks: 'It is, you might say, a museum of a distillery that is still in operation.' It is quirky and proudly independent.

James McEwan, Bruichladdich's former Director of Production, is a legend in the whisky trade. He was trained as a cooper from 1963 at Bowmore Distillery and ran the warehouses there until 1977, when he moved to Glasgow to spend the next seven years as a blender. He returned to Bowmore as Distillery Manager in 1984 and was soon spending much of his time travelling the world, conducting talks and tastings, spreading the whisky gospel, and that of Islay and Scotland. He was persuaded to join the team that took over Bruichladdich in 2000 and retired in 2015.

Bruichladdich became the focus of an intelligence operation by the (American) Defense Threat Reduction Agency, who believed its distilling equipment could also be used to make chemical weapons. The distillery owners learned of this when a helpful American agent informed them that the webcams in the stillhouse, which she was using to monitor the facility, had broken down! Typically, Bruichladdich issued a limited run of commemorative bottles in her honour!

As well as the traditional style of very lightly peated malt, Bruichladdich also produces heavily peated whiskies, including Octomore and Port Charlotte. In 2006 it also produced the strongest malt whisky ever made (at least in modern times), triple-distilled Trestarig and quadruple-distilled Usquebaugh-baul, from an account by Martin Martin's *Description of the Western Isles* (1703). The latter translates as 'perilous whisky' and was filled into cask at 88% vol.

In 2010 a Lomond still from Inverleven Distillery

in Dumbarton was installed at Bruichladdich and, following modifications by Master Distiller Jim McEwan, this commenced distillation of 'The Botanist' Islay dry gin in 2011.

RAW MATERIALS: Bruichladdich reservoir for process water (soft and acidic/peaty), Bruichladdich burn for the cooling water and James Brown's spring (clear water) at Octomore for reduction water. Floor maltings closed 1961; then malt from Port Ellen; now malt from mainland maltings.

PLANT: Cast iron rake-and-plough mash tun (6.2 tonnes). Six Oregon pine washbacks. Two tall plain wash stills (charge 12,000 litres each). Two tall plain spirit stills (charge 7,100 litres each). All indirect fired by steam. Shell-and-tube condensers.

MATURATION: 25% in first-fill and refill ex-sherry hogsheads, 65% in first-fill bourbon barrels, and other types of casks (rum and wine). Eight bonded warehouses on site and another four at Port Charlotte. Most are dunnage, with two racked; 35,000 casks in total. Bottling Hall, named after the Harvey brothers, opened May 2003. Cooperage opened May 2004.

STYLE: Malty, with peaty variations.

MATURE CHARACTER: 'Traditional' (i.e. unpeated) Bruichladdich is fresh, grassy and malty on the nose, with fragments of wild-flower notes. The taste is sweet to start, with cereal and citric notes; dry and short in the finish. Light-bodied.

Bunnahabhain

REGION
Islay

ADDRESS
Near Port Askaig, Islay, Argyll

PHONE
01496 840646

WEBSITE
www.bunnahabhain.com

OWNER
Distell Group Ltd

VISITORS
Welcome. Holiday cottages

CAPACITY
2.7m L.P.A.

HISTORICAL NOTES: William Robertson of Robertson & Baxter, whisky blenders and brokers in Glasgow, founded Bunnahabhain Distillery in 1881, in partnership with the Greenlees Brothers from Campbeltown (owners of the very successful Lorne, Old Parr and Claymore blends, and of Hazelburn Distillery) and with the name Islay Distillers Company Ltd. This became the Highland Distilleries Company Ltd in 1887, when it merged with Glenrothes-Glenlivet.

The distillery's situation is among the most remote in Scotland, and building here was not without its difficulties: two large boilers were blown off the beach, where they waited to be fitted, during the first winter of building. The decision to build here was on account of Robertson & Baxter's close relationship with Bulloch Lade & Company, who were rebuilding Caol Ila at the time, just up the road, and also because of the copious water from Loch Staoinsha and good sea access. The spirit first flowed in January 1883.

The original distillery had one pair of stills. This was doubled in 1963, at which time the floor maltings were removed. Although its capacity was reported to be 3.4 million litres per annum in 1987, it had been closed from 1982 to 2004, and production was down to 750,000 litres by 2002.

Highland Distilleries changed its name to Highland Distillers in 1998 and then took the name of its parent company, The Edrington Group, in 1999. Somewhat surprisingly, the company sold Bunnahabhain to Burn Stewart Distillers in April 2003 for £10 million, and in April 2013 Burn Stewart's holding company, CL World Brands, was bought by the Distell Group of South Africa for £160 million.

CURIOSITIES: As well as the distillery itself, a mile-long road had to be constructed up the steep cliff to join the track to Port Askaig, a pier built out into

Over 250 ships
have foundered
off the coast of
Islay, four of them
within sight of
Bunnahabhain.

the fast-flowing waters of the Sound of Islay, and houses erected for the workforce. The total cost was estimated at £30,000.

When completed, the distillery was much admired. Alfred Barnard, who visited five years after it opened, described it as 'A fine pile of buildings in the form of a square and quite enclosed. Entering by the noble gateway one forms an immediate sense of the compactness and symmetrical construction of the work.' The appearance is little altered today.

Bunnahabhain is Islay's most remote and most northerly distillery. Its make is often described as the mildest Islay malt. This light style was adopted to provide whisky for blending, especially for Cutty Sark and Black Bottle (the latter was also sold to Burn Stewart).

Bunnahabhain is traditionally lightly peated (2–3ppm phenols), but trials with heavily peated malt (35–40ppm) were conducted in 1997, and under Burn Stewart's ownership batches of smoky Bunnahabhain have been produced. The first, called Moine (Gaelic for 'peat'), was offered for sale at 6YO during the Islay Festival 2004.

Since it was first bottled as a single in 1979, Bunnahabhain labels have been illustrated with a sailor behind a ship's wheel and the motto 'Westering Home'. The words come from the well-known Scots song:

Westering Home, with a song in the air,
Light in the eye and it's goodbye to care;
Laughter o' love and a welcoming there,
Isle of my heart my own one.

RAW MATERIALS: Very slightly peated and unusually hard process water from a spring in the Margadale Hills, piped from source so as to keep peat content down. Cooling water from Loch Staoinsha. 10% of malt from Port Ellen Maltings, the rest from Simpson's of Berwick. A portion peated to around 38ppm phenols; the majority unpeated.

PLANT: Large stainless steel rake-and-plough mash tun (12.5–13 tonnes). Six Oregon pine washbacks. Two plain pear-shaped wash stills (charge 16,625 litres each) and two plain onion-shaped spirit stills (charge 9,000–9,600 litres each). All indirect fired by steam. Shell-and-tube condensers.

MATURATION: Mainly refill ex-bourbon hogsheads, with a percentage of 10% first- and second-fill sherry casks. Six dunnage and one racked warehouses on site; approximately 21,000 casks in total. All matured on site.

STYLE: Sweet and fruity, with a smoky batch each year, at 35–40ppm phenols (trialled in 1997; resumed 2003).

MATURE CHARACTER: 'Traditional' Bunnahabhain (i.e. unpeated) is sweet, lightly fruity and faintly maritime on the nose, sometimes with a whiff of peat smoke. The mouth feel is smooth, the taste mild, lightly sweet, then drying, with a hint of smoke. Light- to medium-bodied.

Caledonian Grain Distillery (Redeveloped)

ADDRESS
Haymarket, Edinburgh
LAST OWNER
D.C.L./S.G.D.
CAPACITY
Unknown
CLOSED
1988

Caledonian was capable of producing as much spirit in a day as a small malt distillery produces in a year.

HISTORICAL NOTES: The Caledonian Distillery – for a year it was known as The Edinburgh Distillery – was built in 1855 by Graham Menzies & Company (owners of Sunbury Distillery in Edinburgh, which closed when Caledonian went into full production), near Haymarket railway station. Now only minutes from the centre of the town, in the 1850s it was beyond the city boundary, but it was close to both the Caledonian and the Edinburgh and Glasgow Railways (both of which ran sidings to the distillery). It is believed that it was one of the first distilleries to take advantage of rail access. Process water came from the city mains supply and cooling water from the nearby Forth and Clyde Canal. It was described as 'the model distillery of Europe'.

Caledonian joined the first Lowland distillers' 'Trade Arrangement' the following year (see *Carsebridge*). This allocated the trade between six distillers, in order to control prices; of the six, Menzies & Company had the largest stock of whisky and were allocated 41.5% of the trade (the next largest was John Bald of Carsebridge, with 15%). However, Graham Menzies held back from joining the nascent D.C.L. in 1877: his son had just become a partner in the family business and wanted to retain family ownership. They joined Distillers in 1884.

In common with other grain distilleries, Caledonian operated both pot and patent stills. In the mid 1880s it had three large pot stills (two of which were used for producing 'Irish' whiskey. They were removed about 1900); its Coffey still was the largest in Europe.

In 1966 the distillery came under the ownership of D.C.L. production subsidiary, Scottish Grain Distillers (S.G.D.). Nearly 400 people were employed at that time. The distillery was closed in 1988, the site being sold and partially demolished in 1997. The remaining buildings – 'externally little changed from the 1880s' (Morrice) – were restored and have been converted into flats.

Caledonian was the largest distillery in the world for many years – certainly between 1897 and 1930.

CURIOSITIES: In 1887 Alfred Barnard pronounced Caledonian to be 'the model distillery of Europe' on account of having 'every improvement of machinery and every patent' known to the industry.

Old and New Edinburgh by James Grant (1882): 'The Caledonian Distillery contains the greatest still in Scotland. In order to meet the growing demand for the variety of whisky known as "Irish" the proprietors of Caledonian distillery about 1867 fitted up two large stills of an old pattern, with which they manufacture whisky precisely similar to that which is made in Dublin.'

William Dudgeon Graham Menzies (1857–1943) became Chairman of D.C.L. in 1897, at the same time as the legendary W.H. Ross became General Manager and Secretary. They made a formidable team, and Menzies served for 28 years, handing over chairmanship to Ross in 1925, but remaining on the board for a further 20 years. 'Together Menzies and Ross built D.C.L.' (Charles Craig). He left a personal fortune of £1.4 million (around £40.5 million today).

'Caledonian has always been a one Coffey-still operation, but with a wash pipe diameter of 4⅜th inches and a height of 45 feet, it produces 4,000 litres of alcohol an hour.' (Morrice)

Only once bottled by its owner, in 1986 to commemorate the Commonwealth Games being held in Edinburgh.

Cambus Grain Distillery (Dismantled)

ADDRESS
Tullibody, by Alloa,
Clackmannanshire
LAST OWNER
United Distilleries
CAPACITY
20m L.P.A.
CLOSED
1993

HISTORICAL NOTES: Alloa has been a centre of brewing and distilling, glass-making and textiles for over 200 years. Cambus Distillery was built in 1806 by John Moubray, on the banks of the River Devon close to where it joins the Forth. The name comes from Gaelic *camas*, a creek or small bay, and the site had formerly been a flour mill.

John was succeeded by his son and grandson; pot stills were abandoned in favour of Stein stills in 1826, to be replaced by Coffey stills in 1851. The grandson, Robert Moubray, took Cambus into D.C.L. on its foundation in 1877 (see *Cameronbridge*, *Carsebridge*), and expanded the distillery in 1882 by the acquisition of the Cambus Old Brewery. Alas, most of the buildings were destroyed by fire in September 1914, and for the next 24 years it functioned only as a bonded warehouse and as a maltings for Carsebridge Distillery nearby.

The ruins of the original structure were demolished, except for a small part of the stillhouse, which was incorporated within a new building in 1937. Production was immediately interrupted by the outbreak of World War II, but resumed in 1945. In 1964 the first by-products recovery plant in any grain whisky distillery was installed at Cambus, converted to a dark-grains plant in 1982.

It was closed by U.D. in 1993; the plant has been removed and the buildings are used for cask filling and maturation.

A new cooperage was built on the old distillery site in 2008 – the largest in Scotland, employing around 30 coopers and employing new cask rejuvenation procedures which enable rejuvenation to first-fill quality. This is to cope with the fact that Diageo's capacity has increased by 40m L.P.A. (with the building of Roseisle, and the increase of production at Cardhu, Clynelish, Cameronbridge etc) since 2008.

A former Excise man at the distillery, Philip Snowden, became Chancellor of the Exchequer in the first Labour government of 1947.

CURIOSITIES: In 1905 an English court ruled that 'whisky', whether malt, grain or blended, must be made in a pot still. At appeal the next year, the court was equally divided. On the day this result was announced, 25 June 1906, D.C.L. cheekily launched an advertising campaign for 'Cambus: A Brand of Pure Patent Still Whisky. Not a headache in a gallon'! The 'What is Whisky?' question was only settled by the appointment of a Royal Commission in 1908, which found that patent still spirit could also be called whisky.

The distillery had its own train for taking in cereals and a yeast house for making 'German' yeast. Casks came in and went out by sea from the distillery's own dock. A CO_2 processing plant was built in 1953, a draff drying plant in 1964 and a dark-grains plant in 1982. In 1952 a rectification plant was installed at Cambus to make gin and augment D.C.L.'s production from Wandsworth Distillery. This was an indication of things to come. Today, around 70% of U.K. gin is made in Scotland, mainly at Cameronbridge Distillery.

Cambus drew water from three sources: process water from the Lossburn Reservoir, deep in the Ochil Hills which rise steeply behind the distillery, cooling water from the River Devon and reduction water from Loch Turret.

In 1957/58 the Forth Brewery next door was converted to patent-still whisky production. During its first two years of existence it produced patent-still malt whisky, but then went over to making grain whisky. It was bought by D.C.L. in 1982 and demolished to make way for the dark-grains plant.

Cameronbridge Grain Distillery

ADDRESS
Windygates, Leven, Fife
PHONE
01333 350377
OWNER
Diageo plc
CAPACITY
95m L.P.A. grain whisky;
42m L.P.A. grain neutral
spirit

HISTORICAL NOTES: Cameronbridge is the oldest grain whisky distillery, and the largest. It was also the first distillery in Scotland to produce grain whisky in column stills – prior to this many Lowland distilleries made grain spirit in pot stills.

It was founded in 1824 by John Haig. According to family tradition, in 1822 (aged 20), he was riding with an 'old servant' past the site of the Cameron Mills on the River Leven by Windygates in Fife. For two centuries this mill had enjoyed 'thirled' privileges (local tenants were obliged to have their corn ground there). John turned to the old retainer and said: 'D'ye ken, Sandy. There is money to be made here – aye from whisky.'

He leased the land from its owner, his friend Captain Wemyss, the lease being taken in his father's name since he was a minor, and Cameron Bridge Distillery was licensed in October 1824.

He installed one of the first Stein stills within a year of its invention (by his cousin, Robert Stein, in 1828), paying the Steins 1d per gallon royalty, to make 'malt aqua'. Two years later he switched to the more efficient Coffey still. This led to a glut of grain whisky, and as early as the mid 1830s Haig was trying to interest the Eastern Lowland distillers in a scheme to regulate prices.

This came to fruition in 1865 with the foundation of the Scotch Distillers Association, a trade agreement between eight grain distilleries to divide up the market according to their production capacity, and to fix prices and conditions of sale. This was the forerunner of D.C.L., established in April 1877 as a combination of grain whisky distillers (Port Dundas, Carsebridge, Cameronbridge, Glenochil, Cambus and Kirkliston) which together controlled 75% of grain whisky production. The nominal capital was £2 million, and its stated purpose was 'to secure the benefits of combined experience and the advantage (which manufacturing and trading on a large scale alone

In August 1877 the River Leven burst its banks and the distillery was 'completely surrounded by water, and a rapid stream ran through the centre court, drowning out the boiler fires ... The main entrance gate gave way under pressure ... and with its demolishment empty casks rolled away on the turbulent surface of the waters ... Much difficulty was experienced in saving the lives of horses and pigs.'

can command) of reduced expenses and increased profits'. John's son, Hugh Veitch Haig, took over the management of Cameronbridge on his father's death in 1878.

Cameronbridge was expanded in 1903 when Drumcaldie malt distillery, nearby, was bought by D.C.L. and absorbed. Until the 1920s it had a combination of Coffey, Stein and pot stills: the last two were removed by 1930.

In 1989 the distillation of Gordon's, Booth's and Tanqueray London Dry Gins was transferred to Cameronbridge from Wandsworth. Today, around 70% of U.K. gin is made in Scotland, mainly at Cameronbridge.

Diageo have spent £70 million since 2010; expanding the site included a new tun room, new fermenters, three new mash tuns and £69 million on an effluent treatment plant, which will supply energy (methane) to run the distillery.

CURIOSITIES: An earlier distillery at Cameronbridge had been worked by John Edington and Robert Haig from about 1813 to 1817. John himself had been born in Cameron House in 1802.

John Haig was a scion of the great Haig distilling dynasty. His father was William Haig, licensee (after 1795) of Kincaple Distillery, close to St Andrews, and founder (in 1810) of Seggie Distillery at Guardbridge (see Eden Mill) – also Lord Provost of St Andrews for twelve years. Young John attended St Andrews University, where he won a silver medal for mathematics, and then served an apprenticeship at Seggie Distillery.

All his uncles were distillers: James at Canonmills, and later Lochrin, Edinburgh – the spokesman for the whisky industry from the 1790s; John at Bonnington Distillery, Edinburgh; George, who had an interest in Inverkeithing Distillery; Robert founded Dodderbank in Dublin, and later took over Seggie; and Andrew, who took over Kilbagie Distillery at Alloa.

When Alfred Barnard visited Cameronbridge in the 1880s he found 'two Coffey's patent stills, handsome machines with their cooper tubes brilliantly polished . . . two of Stein's Patent Steam Stills . . . an old Pot Still.' He concluded: 'The whisky made here is said to have no rival in the world. There are several kinds manufactured. First "Grain Whisky", second "Pot Still Irish", third "Silent Malt" and fourth "Flavoured Malt".'

Cameron Brig Pure Single Grain has long been the only single grain whisky widely available. In July 2013, it was replaced by Haig Club Single Grain Whisky @ 40% vol with no age statement, a mix of refill ex-bourbon casks and rejuvenated European oak casks, designed as a 'recruitment whisky' – and promoted by David Beckham!

RAW MATERIALS: Wheat from the East Coast of Scotland and from England; Wanderhaufen malting system on site until 1997, now dried malt from Burghead. Process water from bore-holes on site.

PLANT: Three Coffey stills, and one Patent still with nine columns for making neutral spirit, for gin and vodka.

MATURATION: Mainly first-fill and refill ex-bourbon casks; matured at Leven, nearby, and Blackgrange, Alloa, in racked warehouses.

STYLE: Light and clean, creamy, oily texture; fruity and flavourful.

MATURE CHARACTER: Fresh, with pear drops (acetone, natural turpentine), on a base of hard toffee. Clean and sweet, with some caramel and a short finish.

Caol Ila

REGION
Islay

ADDRESS
Port Askaig, Isle of Islay,
Argyll

PHONE
01496 302760

WEBSITE
www.malts.com

OWNER
Diageo plc

VISITORS
Yes

CAPACITY
6.5m L.P.A.

*During the period
1970–74, when
the distillery was
rebuilt, peating
levels were raised
at Brora Distillery
to make up for the
potential shortfall
of smoky whisky.*

HISTORICAL NOTES: The distillery takes its name
(pronounced 'Cull-eela') from the strait between
the islands of Islay and Jura, the Sound of Islay.
The first distillery on this site – a small bay just
north of the ferry point at Port Askaig – was built
in 1846 by Hector Henderson, who took advantage
of copious water from Loch nam Ban. The site had
previously been used for washing lead ore. It enjoys
splendid views across to Jura. A number of houses
were built on the hillside above the distillery, with a
shop and Mission Hall for the employees.

In 1863 Caol Ila was bought by the Glasgow
blending firm Bulloch Lade & Company, who built
a pier capable of withstanding the 12-foot tide-fall
and the strong currents of the Sound, which allowed
small cargo ships (puffers) to supply coal and barley
and uplift whisky.

In 1927 D.C.L. acquired a controlling interest in
Bulloch Lade, and ownership went to S.M.D. in 1930.
They bought their own puffer, the *Pibroch*, to service
their three Islay distilleries.

In 1972 the original distillery was demolished
(apart from the large, three-storey warehouse,
which is still in use) and replaced by a larger
and more efficient building in S.M.D.'s so-called
'Waterloo Street' style. It resumed production in
1974, with six stills (where formerly it had two).

During 2011/12 Diageo added a further two
stills and spent £3.5 million on the site to increase
capacity by 700,000 L.P.A.

CURIOSITIES: The 'Waterloo Street' style was named
after the Glasgow address of S.M.D.'s engineering
department. The design was for a six-still distillery
and specified that the outside wall of the stillhouse
be of glass, with windows that could open, making
them light-filled and airy – and in Caol Ila's case
providing a stunning view to the Isle of Jura.

The mashhouse and tun-room were arranged in
such a way as to make the best use of gravity, thus

Billy Stichell, who retired in 2014, was the fourth generation of his family to have worked here.

saving unnecessary pumps, and the overall plan was both efficient and aesthetically pleasing, while also being pleasant to work in. The design was the brainchild of Dr Charlie Potts, S.M.D.'s Chief Engineer, and was applied in the following distilleries: Balmenach (1962), Caol Ila (1974), Clynelish (1968), Craigellachie (1965), Glendullan (1972), Glen Ord (1966), Glentauchers (1966), Linkwood (1970), Mannochmore (1971), Royal Brackla (1964/65) and Teaninich (1970). Although the style of Caol Ila is famously peated, each year a batch of unpeated malt is made, named Caol Ila Highland.

RAW MATERIALS: Heavily peated malt (30–35ppm phenols) from Port Ellen Maltings, and, for around four months of the year, unpeated malt from Port Ellen. Water from Loch nam Ban.

PLANT: Cast iron semi-Lauter mash tun (11.5 tonnes). Eight larch washbacks. Three plain wash stills (35,340 litres charge), three plain spirit stills (29,550 litres charge). Indirect fired by steam throughout. Shell-and-tube condensers.

MATURATION: Refill ex-bourbon hogsheads. Mainly matured in the Central Belt, some on site in the original warehouse.

STYLE: Peaty, but with a lighter body than its sister Lagavulin.

MATURE CHARACTER: Sweet and lightly fruity on the nose, with smoked ham or smoked cheese, and some seaweed. The taste is sweet, with fragrant smoke and antiseptic cream. Medium length.

Caperdonich Demolished

REGION
Speyside

ADDRESS
Rothes, Moray

LAST OWNER
Chivas Brothers

CAPACITY
2.1m L.P.A.

The 'whisky pipe' which connected Caperdonich to Glen Grant was a popular source of free spirit among the 'free spirits' of Rothes!

HISTORICAL NOTES: At the height of the Whisky Boom, the demand for Glen Grant malt was such that the distillery's owner, Major James Grant, built another distillery 'across the road', which he imaginatively named Glen Grant Number Two (see *Glen Grant*). It was designed to augment supplies of Glen Grant, the make of each being considered the same, and there was a pipe connecting the two. However, it is said that from the outset, the product of Number Two was quite different from Number One.

It opened in 1898, two years before the Boom turned to bust, and the distillery was closed in 1902, the floor maltings and two kilns being used to supply malt to Glen Grant Distillery, and warehouses used to store materials.

It remained silent until 1965, when it was rebuilt by The Glenlivet Distillers Ltd, expanded in 1967 to four stills and renamed Caperdonich after the well which supplies its reducing water. In 1977 it was acquired by Seagram when they bought Glenlivet Distillers, and by Pernod Ricard when that company bought Seagram's whisky interests in 2001. The following year Chivas Brothers, Pernod Ricard's operating company, mothballed the distillery. In 2012, The Forsyth Group (the family owned coppersmiths in Rothes founded in the 1890s and were responsible for making most of the new stills, receiver tanks and piping for many of today's new distilleries around the world) bought the Caperdonich site, demolished the distillery buildings and now use the site in conjunction with their fabrication business down the road.

CURIOSITIES: The stills installed in 1965/67 contained large parts of the original Glen Grant No. 2 stills, identifiable by their riveted seams, but they have now been replaced. Although it uses the same malt and water, Caperdonich's make is lighter.

Glen Grant No. 2 was connected to its sister by a pipe, so new make was pumped up the hill to be filled into cask at the larger distillery.

Cardhu

REGION
Speyside

ADDRESS
Knockando, Aberlour,
Moray

PHONE
01479 874635

WEBSITE
www.discovering-
distilleries.com

OWNER
Diageo plc

VISITORS
Visitor centre – the 'brand
home' of Johnnie Walker,
with memorabilia

CAPACITY
3.4m L.P.A.

HISTORICAL NOTES: Like many others, Cardhu (until 1981 known as 'Cardow', Anglicised to 'Cardoo') was a farm with a still. John Cumming, the son of a hill-farmer and grazier, became tenant here in 1811, and soon turned his hand to illicit whisky-making. The location, in hilly country to the north of the Spey, was, in those days, remote. He took out a licence in 1824; much of his make was taken by horse and cart to Burghead and shipped to Leith.

John died in 1846 and was succeeded by his son, Lewis Cumming, and Lewis by his wife, Elizabeth – a faded photograph of whom is in the distillery. At this time the output of the distillery was 240 gallons a week (623 litres).

In 1884 she rebuilt the distillery, selling the old stills to William Grant, who was building Glenfiddich Distillery at the time. By 1888 her whisky was being sold as a single malt in London, proudly claiming to be the only whisky from Speyside that did not need to affix the name Glenlivet to its own.

Elizabeth Cumming sold to John Walker & Sons in 1893, on condition that her son, John, joined the Board of the company. Her grandson, Ronald, would ultimately rise to be Chairman of both Walkers and D.C.L., and a Knight of the Realm.

In 1897 the distillery was expanded to four stills, and in 1925 Walker's joined D.C.L. In 1965, noting the success of Glenfiddich, the D.C.L.'s Home Trade Committee (backed by Sir Ronald Cumming) investigated possible malts for promotion as singles, and recommended Cardow/Cardhu, Aultmore and Linkwood. A budget of £15,000 was put behind a release of Cardhu @ 8YO, although the management committee were very nervous about eroding stock for blending purposes, and also concerned that advertising single malt as 'something special' was 'not perhaps knocking standard brands'. 'The experiment' was put on hold: Cardhu could continue to be sold in the home market, but no more was to be spent on advertising after 1967.

In the early 1980s the company took the unusual step (for them) of allowing journalists and non-whisky trade V.I.P.s to visit the distillery.

Since 2011, Cardhu has expanded production from 2.3 million to 3.4 million L.P.A. by operating 24/7.

CURIOSITIES: John Cumming's wife, Helen, aided her husband and others in their illicit distilling ventures by boldly offering accommodation in the farmhouse to visiting excisemen, there being no inn for miles around. As soon as they were safely at table, she hoisted a red flag over the barn at the back as a warning.

Barnard had a high regard for the make from Cardow, describing it as 'of the thickest and richest description and admirably suited to blending purposes'. Charles Mackinlay & Co in Leith were agents, and it commanded a premium price.

Alfred Barnard visited Cardow shortly before Elizabeth Cumming rebuilt the distillery. He reported that the buildings were 'of the most straggling and primitive description and although water power existed, a great part of the work was done by manual labour. It is wonderful how long this state of things existed, considering the successful business that was carried on for so many years.'

In 1924 a trade journal reported that the 'two larger stills' at Cardhu were fired by 'oil and steam pressure on the jet system' – i.e. direct fired by oil – while the two smaller stills continued to be heated by coal fires. This was an experiment far ahead of its time. The experiment was abandoned after two years, on account of the cost of oil, not the quality of spirit.

In many ways, Sir Alexander Walker was an innovator, but in others he was very conservative. He subscribed to the widely held belief that it was unlucky to interfere

Elizabeth Cumming was a formidable character who became known as 'The Queen of the Whisky Trade'.

with spiders in the tun room, as being beneficial to fermentation. On his orders they were a protected species, so when a new brewer at Cardhu unwittingly cleaned and repainted the room . . . 'Sir Alexander's anger, when he discovered this act of impiety, was long remembered' (Brian Spiller).

Until the current recession, Cardhu was so popular in Spain that there was not enough to supply the demand. In 2002 Diageo sought to solve the problem by introducing a 'pure malt' (a blend of malts – in this case only two) version of Cardhu. It was named Cardhu and offered in the same style of bottle, with a very similar label, although the diffi rence between 'pure' and 'single' malt was explained on the carton, and it was planned to revert to the original name Cardow for the single malt. The move caused an uproar in the whisky industry, unprecedented in modern times, and led to a tightening of the definitions of malt whisky (the word 'pure' was banned, for example, as being misleading to consumers). Cardhu Pure Malt was withdrawn.

RAW MATERIALS: Unpeated malt from Burghead Maltings. Soft water from springs on Mannoch Hill and from the Lynne Burn, both collected in a dam near the distillery.

PLANT: Full-Lauter mash tun (seven tonnes). Eight larch washbacks. Three plain wash stills (18,000 litres charge); three plain spirit stills (11,000 litres charge). Indirect fired since 1971. Shell-and-tube condensers.

MATURATION: Refill ex-bourbon hogsheads.

STYLE: Floral.

MATURE CHARACTER: Fragrant and floral (Parma Violets, dried rose petals), and Speyside-fruity (pear drops, fresh apples) on the nose. Sweet and fresh to taste. Light-bodied.

Carsebridge Grain Distillery (Redeveloped)

ADDRESS
Alloa, Clackmannanshire
OWNER
Diageo plc
CAPACITY
Unknown
CLOSED
1983

Carsebridge once had its own fire brigade, employing 40 firemen.

HISTORICAL NOTES: Clackmannan was the cradle of brewing and distilling in Scotland during the late eighteenth century. Alloa stands in the shadow of the Ochil Hills, at the junction of the Stirling Plain and the fertile lowlands of Fife; it was close to some of the earliest coal pits in Scotland, and had access to the Forth for the import of coal and grain from East Lothian.

Here were found 'the largest manufacturing undertakings of any kind to emerge during the first decade of the industrial revolution in Scotland' (Michael Moss) – Kilbagie and Kennetpans Distilleries, owned by the Stein family, and associated trades, such as coppersmiths, glass works and cooperages (all of which still exist).

Carsebridge was built in 1799 by John Bald, who left the running of it to his son, Robert. In 1845 it passed to Robert's brother, John. It began as a pot-still malt whisky distillery, but switched to patent stills and grain whisky production in 1860.

John Bald II (described as 'the politic Bald') was a leading light in the move towards promoting the collective interests of the Lowland distillers. This first took the form of a 'Trade Arrangement for one year', signed in 1856 by the owners of Caledonian, Cambus, Carsebridge, Glenochil, Haddington and Seggie Distilleries. This was extended by a second Trade Arrangement in 1865, when Cameronbridge replaced Seggie and Port Dundas replaced Haddington; later the same year Yoker and Adelphi Distilleries joined.

All this was but a prelude to the formation, in April 1877, of D.C.L. (see *Cameronbridge*).

Carsebridge was transferred to the ownership of D.C.L. subsidiary S.G.D. in 1966, at which time it was the largest grain distillery in the group, employing 300 people, with three Coffey stills and a large dark-grains plant. It was closed and dismantled in 1983 and the site is now used by Diageo's Spirit Supply, Scotland division, which is responsible for all the company's spirits production. The site also houses the company's principal cooperage and 'bodega' – the latter for seasoning casks with sherry or other wines.

CURIOSITIES: The first steam engine in Scotland was installed in the 1770s at Kennetpans Distillery, while Kilbagie had the first steam-powered threshing mill. One of the earliest railway lines in Scotland connected the two distilleries, so that goods could easily be moved from Kilbagie to the wharf at Kennetpans, a mile distant.

Clynelish

REGION
Highland (North)

ADDRESS
Brora, Sutherland

PHONE
01408 623003

WEBSITE
www.malts.com

OWNER
Diageo plc

VISITORS
Visitor centre and shop

CAPACITY
4.8m L.P.A.

HISTORICAL NOTES: The present-day Clynelish Distillery was built in 1967/68, close to the original Clynelish Distillery, which was renamed Brora (see this entry for the early history), as part of D.C.L.'s expansion policy at that time. Architecturally, it bears a resemblance to Caol Ila, Mannochmore and Craigellachie – all of which were rebuilt at around the same time, in the so-called 'Waterloo Street' style (see *Caol Ila*).

The new Clynelish Distillery was equipped with six stills, all steam heated from an oil-fired burner. The spirit is highly sought after (not least for the Johnnie Walker blends), and since 2011 output was increased from 3.36 million LPA to 4.8 million LPA by operating 24/7. In January 2014 Diageo announced plans to increase capacity to over 9 million LPA, but later that year this plan was put on hold. During 2016 there was a major upgrade of production plant, however, including a new mashtun.

CURIOSITIES: The old Clynelish Distillery was officially closed in 1967, but continued in production. From 1972 to 1977 it produced heavily peated whisky, to make up for the loss of peaty spirit from Caol Ila Distillery, which was being rebuilt at the time. Its name was only changed to Brora in 1975, so for seven years 'Clynelish' was being made at both distilleries.

Clynelish is unusual in having spirit stills larger than its wash stills. Its spirit style is uniquely 'waxy', and this flavour comes through in the mature whisky. Until recently it was a mystery where this prized waxiness came from. A full investigation was prompted by a temporary change in spirit character and it was discovered that the characteristic was contributed by deposits of greasy matter which built up in the receivers and piping. Its temporary absence was on account of the piping having been

Clynemilton Burn has supplied the town of Brora since 1819 and is said to contain particles of gold; certainly gold can be panned for in neighbouring rivers.

cleaned when a receiver was replaced. Such are the mysteries of where flavour comes from in malt whisky!

RAW MATERIALS: Unpeated malt from Glen Ord. Soft water from the Clynemilton Burn.

PLANT: Full-Lauter mash tun, with copper canopy (12.5 tonnes). Eight larch washbacks. Three boil-pot wash stills (17,000 litres charge); three boil-pot spirit stills (20,000 litres charge). All indirect fired. Shell-and-tube condensers.

MATURATION: Mainly ex-bourbon refill casks, with some ex-sherry. Two dunnage warehouses on site (7,000 casks). Most matured in the Central Belt.

STYLE: Waxy, with heather notes.

MATURE CHARACTER: An outstanding example of the 'Highland' style, Clynelish is scented with heather flowers and moorland herbs, candlewax and fragrant smoke. The texture in the mouth is waxy, teeth-coating; the taste creamy, lightly fruity and spicy, with tobacco notes. Medium-bodied. Complex.

Coleburn Redeveloped

REGION
Speyside

ADDRESS
Longmorn, by Elgin,
Moray

PHONE
07724 045221

WEBSITE
www.thewhiskyhotel.
com

OWNER
Aceo Ltd

VISITORS
Visitor centre and shop

CLOSED
1985

*The name Coleburn
might be translated
as 'Charcoal Burn':
the district was
popular for charcoal
making in the past.*

HISTORICAL NOTES: Coleburn was built in 1896 by John Robertson & Son Ltd, blenders in Dundee, and designed by Charles Doig. It has an attractive situation within the Glen of Rothes, chosen to take advantage of a branch-line of the Great North of Scotland Railway, and the distillery had its own small station and siding (closed in 1966). The name recollects that charcoal was made nearby.

The distillery closed in 1913 and was sold three years later to Clynelish Distillery Company Ltd (itself owned by D.C.L., John Walker & Sons and John Risk), passing to full D.C.L. ownership in 1925. It was licensed to J. & G. Stewart, and the make was a key filling in the Usher's blends. It closed in 1985. The attractive site was sold to Dale and Mark Winchester in 2004 for development.

Their intention was to develop the 14-acre site into a luxury 'Whisky Hotel', with associated leisure facilities, but in November 2013 they sold part of the property to whisky brokers Aceo Ltd, including the extensive traditional dunnage warehouses, the filling store and receiver room. The Winchester brothers have retained the old maltings and still hope to realise their hotel dream.

Aceo already owns thousands of casks, which have been moved into the bond, which was licensed by HMRC in 2014, along with the casks they acquired from Murray McDavid. Remaining space in the warehouses will be rented out for cask maturation.

It is the company's firm intention to resume production at Coleburn. Costings are being done at the moment and an order for stills from Forsyth of Rothes is in process.

CURIOSITIES: As experienced whisky brokers, Aceo are acutely aware of the importance of maturation in good wood and the right atmospheric conditions. They offer a comprehensive cask management service – regular monitoring and maturation reports, cask repair and certification – and also hold a large number of first-fill ex-wine, port, rum, sherry and bourbon casks, available to customers for finishing purposes. This is the first time such services have been offered.

Convalmore Dismantled

REGION
Speyside

ADDRESS
Dufftown, Moray

LAST OWNER
William Grant & Sons

CLOSED
1985

HISTORICAL NOTES: Convalmore was the fourth distillery to be built in Dufftown, next door to Balvenie and Glenfiddich (the buildings, now used as storage and warehousing, are owned by William Grant & Sons, owners of the adjacent distilleries). Building commenced in 1893 and production started the following year.

The owner was the Convalmore-Glenlivet Distillery Company, a group of Glasgow blenders, with Peter Dawson (a Dufftown man, and a well-known wholesale whisky merchant) as Managing Director. In spite of this ready market for its make, the company failed in 1905 and was bought by W. P. Lowrie & Company, blenders in Glasgow and suppliers of whisky to James Buchanan & Company, which acquired Lowrie's in 1907. Many of the buildings were destroyed by fire in 1909, but they were rebuilt the following year (incorporating a continuous still, but this was abandoned in 1915).

Buchanan's joined D.C.L. in 1925, and ownership went to S.M.D. in 1930. Convalmore was closed in 1985, and the buildings sold to William Grant & Sons in 1990.

CURIOSITIES: The distillery name comes from the Conval Hills, from which the process water was drawn. Cooling water came from the River Fiddich. During the fire of 1909, the heat was so intense that the distillery's hoses could not be connected and workers had to resort to carrying buckets of water up from the Fiddich. 'While it was at its height the flames rose to between 30 and 40 feet high . . . To add to other discomforts, snow commenced to fall, and the effect of the burning building on the white landscape provided a striking picture.'

Cragganmore

REGION
Speyside

ADDRESS
Ballindalloch, Moray

PHONE
01479 874700

WEBSITE
www.malts.com

OWNER
Diageo plc

VISITORS
Yes

CAPACITY
2.2m L.P.A.

HISTORICAL NOTES: A distillery at Ballindalloch, deep in Speyside, was only made possible by the opening of the Strathspey railway in 1863. Its founder, John Smith, was acknowledged to be one of the most experienced distillers of his day, having formerly managed Macallan, Glenlivet, Glenfarclas and Wishaw Distilleries; he was also a keen railway enthusiast and a moving force behind the building of the Speyside line.

The site he chose for his distillery was on Ballindalloch Estate, and he built it with the support of Sir George Macpherson Grant, the laird, whose family had owned Ballindalloch Castle since the fifteenth century (*see Ballindalloch*).

The distillery opened in 1869. On John's death in 1886, it passed to his brother, and then to his son, Gordon Smith, when he came of age in 1893. He refurbished it in 1901, employing Charles Doig, the famous distillery architect; between his death in 1912 and its sale in 1923, Cragganmore was managed by his widow Mary Jane.

The new owner was a partnership between Macpherson Grant and White Horse Distillers; the latter's 50% passed to D.C.L. in 1927, and in 1965 S.M.D. bought out the remaining share, having doubled the capacity (to four stills) the previous year.

CURIOSITIES: For reasons unknown, John Smith designed his stills with flat tops, rather than the usual swan necks. This may increase reflux, the spirit vapour condensing on the flat surface and dripping down to be redistilled. Any lightness of body this might contribute is countered by the use of worm tubs. The result is a medium-bodied Speyside of unusually complex character.

John Smith himself was a very large man – he weighed 308 lbs (140 kg) – and was too wide to enter a railway carriage. Accordingly, he was obliged to travel in the guard's van.

Cragganmore is ranked Top Class by blenders. In 1988 it was chosen by its owners to represent the Speyside style in their Classic Malts series.

Cragganmore is a very attractive distillery, compact and neat. It is built around a courtyard, one side of which houses the 'Club Room' (used for entertaining guests), which is furnished in Edwardian style, with memorabilia including John Smith's desk and (large!) chair.

RAW MATERIALS: Lightly peated malt from Roseisle. Hard process water from a spring on Craggan More Hill via the Craggan Burn; cooling water from the River Spey.

PLANT: Full-Lauter mash tun with copper canopy (seven tonnes). Six Oregon pine washbacks. Two lamp-glass wash stills (18,500 litres charge); two boil-ball spirit stills (6,000 litres charge), with unusual flat tops, all indirect fired. Unusual rectangular worm tubs, with worms from both wash and spirit stills in the same tub.

MATURATION: A mix of refill ex-bourbon and ex-sherry casks.

STYLE: Big, rich and meaty.

MATURE CHARACTER: A multi-layered nose of polished leather and saddle-soap, green bananas, tobacco, nuts, dried fruits and herbs. The taste is dryish overall, with walnuts, hard toffee and dried fruits. Medium-bodied.

Craigellachie

REGION
Speyside

ADDRESS
Craigellachie, Aberlour, Moray

PHONE
01340 872971

WEBSITE
www.craigellachie.com

OWNER
John Dewar & Sons Ltd (Bacardi)

VISITORS
By appointment

CAPACITY
4,1111 P Å

HISTORICAL NOTES: Craigellachie Distillery was substantially rebuilt and expanded from two stills to four in 1964/65 by D.C.L. (its then owner), to the 'Waterloo Street' design (see *Caol Ila*).

All that remains of the earlier distillery are parts of the warehouses. The original was designed by Charles Doig and built in 1891 (but did not start production until 1898) for a consortium of blenders and merchants, led by Peter Mackie of White Horse and Alexander Edward, owner of Benrinnes and Aultmore Distilleries. The latter pulled out in 1900, and Mackie & Company took over complete ownership in 1916. 'From that date (1900) onwards the annual general meeting of the company provided the Chairman, Peter Mackie, with a platform for strongly held opinions on the state of the whisky industry, the nation and the British Empire.' (Brian Spiller)

Sir Peter Mackie died in 1924; the company changed its name to White Horse Distillers and joined D.C.L. three years later.

In 1998 U.D.V. (successor to D.C.L.) was obliged to sell Craigellachie, with three other distilleries, to Bacardi along with the Dewar's brands. In 2014 three new expressions for domestic markets were released (at 13, 17 and 23 years old) and one for duty free at 19 years.

CURIOSITIES: Peter Mackie was one of the leading characters of the whisky industry. Among his eccentricities was the invention of a 'power flour' called B.B.M. (Bran, Bone and Muscle), mixed to a secret recipe under the boardroom at Craigellachie, which all employees were required to use at home. Staff were also required to tend their gardens carefully, and an annual prize was offered, after a tour of inspection by the Directors. He was an ardent Conservative, but was made a baronet in 1920 for his war work by a Liberal prime minister.

Peter Mackie was described as 'one third genius, one third megalomaniac and one third eccentric'.

The distillery was lit by paraffin lamps until 1948, and a water wheel was used to drive the wash still rummager until 1964, when the distillery was completely rebuilt.

RAW MATERIALS: Floor maltings until 1964; thereafter lightly peated malt from Burghead. Soft process water from springs on Little Conval Hill; cooling water from the River Fiddich.

PLANT: Full-Lauter mash tun (ten tonnes) installed 2001. Eight larch washbacks. Two plain wash stills (22,730 litres charge); two plain spirit stills (22,730 litres charge). All indirect fired since 1972. Worm tubs.

MATURATION: Mainly refill ex-bourbon hogsheads, some ex-sherry butts; matured in the Central Belt.

STYLE: A big-bodied and very slightly smoky Speyside style.

MATURE CHARACTER: Craigellachie is unusual as a Speyside in adding a thread of smoke to the familiar floral-fruity characteristic of the region. Sweet and creamy in the mouth, with citric notes, light acidity and a trace of smoke in the finish. Full-bodied.

Daftmill

REGION
Lowland

ADDRESS
Daftmill Farm, Cupar, Fife

PHONE
01337 830303

WEBSITE
www.daftmill.com

OWNER
Francis and Ian Cuthbert

VISITORS
By appointment

CAPACITY
c. 65,000 L.P.A.

The Daft Burn, which gave its name to the mill, is so called because it appears to flow uphill.

HISTORICAL NOTES: Daftmill is among the most attractive distilleries in Scotland. It is a neat, sensitively converted meal mill, built around three sides of a square, with the glass-fronted stillhouse at the base, a warehouse to the right and mashhouse/tun room in the left-hand range. The mill building itself dates from the late seventeenth to nineteenth centuries (the date on the gablestone reads 1809).

The conversion, which was done between 2003 and 2005, was the work of brothers Ian and Francis Cuthbert, whose family has been farming the land hereabouts for six generations. They also own a gravel quarry nearby, income from which provided the money to convert the mill.

CURIOSITIES: All the work to convert the mill was done by men from within a five-mile radius of Daftmill Farm, except for the stills and mash tun, which were made by Forsyth of Rothes.

The brothers have their own barley malted by Crisp of Alloa. At the time of writing, Daftmill has not been released as a single malt. When asked when it will be released, Francis Cuthbert says simply: 'When it's ready'!

RAW MATERIALS: Unpeated malt from Crisp, Alloa. Process and cooling water (hard) from a spring on site.

PLANT: Semi-Lauter mash tun with copper canopy (one tonne). Two stainless steel washbacks. One plain wash still (2,500 litres charge); one plain spirit still (1,500 litres charge); both indirect fired. Shell-and-tube condensers.

MATURATION: A mix of first-fill ex-bourbon (from Heaven Hill Distillery) and ex-sherry casks.

STYLE: Light and fruity, with cereal notes.

MATURE CHARACTER: Light cereal, toffee and citrus notes, with the latter being replaced by dry-fruit notes when matured in sherry casks.

Dailuaine

REGION
Speyside

ADDRESS
Carron, Moray

PHONE
01340 872500

WEBSITE
www.malts.com

OWNER
Diageo plc

VISITORS
Visitor centre and shop

CAPACITY
5.2m L.P.A.

HISTORICAL NOTES: The distillery was built in 1851 by William Mackenzie, in a small wooded glen a mile off the main road, at the other (northern) end of Ballindalloch Estate from Cragganmore. Access to market was given a terrific boost twelve years later by the arrival of the Speyside Railway on the opposite bank of the river, at Carron, connected by a road bridge.

The distillery was rebuilt in 1884, and when Alfred Barnard visited in 1887 he noted: 'Within the last few years, nearly the whole of the distillery has been rebuilt on a larger and more modern style, and the work now contains all the latest improvements in the art of distilling.' It was now one of the largest Highland distilleries.

Mackenzie & Company became a limited company in 1890, and merged with Talisker to become Dailuaine-Talisker Distilleries Ltd in 1898. By this time the business was in the hands of Mackenzie's son, Thomas, who had built Imperial Distillery the previous year. When Thomas died without heirs in 1915 Dailuaine-Talisker was bought by its principal customers, Walker's, Dewar's, W.P. Lowrie and D.C.L. A fire destroyed much of the property in 1917, and it was rebuilt.

A major reconstruction took place in 1959/60, when the floor maltings were converted into a Saladin box system, the number of stills increased from four to six and mechanical coal-stoking was introduced (the distillery went over to indirect firing in 1970). A dark-grains plant was also installed at this time. Malting on site remained until 1983.

Always designed to be a blending whisky, Daluaine was not bottled as a single by its owners until 1991.

CURIOSITIES: The pagoda roof, which has since become the *leitmotif* of a malt whisky distillery, was destroyed by the fire in 1917. There was another

Charles Doig of Elgin, the pre-eminent distillery architect of his day, installed his first pagoda at Dailuaine in 1889.

fire in 1959, and this prompted the refurbishment referred to above.

A 0–4–0 saddle locomotive, built by Barclay of Kilmarnock, was bought in 1897 to run goods down to Carron Station. It operated until 1939, when it was replaced by another engine from the same works, named Dailuaine No. 1 – described by a former driver as 'truly a joy to behold', with its bright paintwork and polished brasswork. When the Speyside line was closed in 1967, it was donated to the Railway Museum in Aviemore. It was later returned to U.D. and is now on display at Aberfeldy Distillery.

RAW MATERIALS: Floor maltings until 1959 then Saladin boxes until 1983; thereafter unpeated malt from Burghead. Soft process water from Balliemullich Burn, flowing down from Ben Rinnes; cooling water from the Green Burn or the River Spey.

PLANT: Full-Lauter mash tun (11.5 tonnes). Eight larch washbacks. Three lamp-glass wash stills (19,000 litres charge); three plain spirit stills (21,000 litres charge). All indirect fired. Shell-and-tube condensers, unusually two of them made from stainless steel – one wash and one low wines – making for heavy spirit.

MATURATION: Mainly refill ex-bourbon hogsheads, with some ex-sherry butts; all spirit now tankered for maturation in the Central Belt.

STYLE: Full-bodied, rich and sulphury.

MATURE CHARACTER: The style of the new-make spirit makes it eminently suitable for sherry-wood maturation, and these flavours emerge in the mature whisky. The nose is redolent of dried fruits and fruit cake, moist with sherry. A hint of rubber immediately after water is added. A thick, unctuous texture in the mouth, with a taste that starts sweet in the mouth and finishes slightly tannic, with dark chocolate. Full-bodied.

Da

Dallas Dhu Museum

REGION
Speyside

ADDRESS
Mannachie Road, Forres,
Moray

PHONE
01309 676548

WEBSITE
www.historic-scotland.
gov.uk

OWNER
Historic Scotland

CLOSED
1983

HISTORICAL NOTES: This is the only distillery 'preserved in aspic' as a museum, an admirable time warp back to the 1950s, and a chance to see (albeit in a somewhat sanitised setting) what distilleries used to be like before modern changes took place.

The ubiquitous Alexander Edward (see *Benromach* etc.) granted the site on his estate to the well-known Glasgow blenders Wright & Greig Ltd. They instructed Charles Doig in 1898, and Dallas Dhu (originally named Dallasmore) went into production the following June.

Ownership passed to J.P. O'Brien & Company (distillers in Glasgow) in 1919, and then to a consortium of English brewers called Benmore Distilleries Ltd. The latter was acquired by D.C.L. in 1929. Closed during the 1930s, the stillhouse was destroyed by fire in 1939, and then the distillery was closed for the duration of World War II, resuming production in 1947. In 1963 the coal-fired stills were equipped with mechanical stokers, and converted to internal heating by steam in 1971.

Dallas Dhu was terminally closed in May 1983, and sold to Historic Scotland in 1986.

CURIOSITIES: Dallas Dhu derives from the Gaelic 'Dail eas dubh' ('the field by the black/dark waterfall').

Wright & Greig's key brand was Roderick Dhu, named after a character in Walter Scott's *The Lady of the Lake*. In the 1880s and 1890s this was a big export Scotch, particularly in India and the Antipodes. Dallas Dhu was built to supply fillings for it.

The parish and ancient church of Saint Michael were granted in 1279 to one William de Ripley, who changed his name to 'de Dallas'. Dallas, Texas, was named after one of his descendants, the U.S. Vice-President George M. Dallas, in 1845.

Dalmore

REGION
Highland (North)

ADDRESS
Alness, Ross and
Cromarty

PHONE
01349 882362

WEBSITE
www.thedalmore.com

OWNER
Whyte & Mackay Ltd

VISITORS
New visitor centre
opened in 2004; cottages
available

CAPACITY
4m L.P.A.

HISTORICAL NOTES: Alexander Matheson, who founded Dalmore in 1839, was a partner in the famous Far East trading company, Jardine Matheson, established by his uncle Sir James Matheson, which soon became the largest British trading company in the Far East. During the 1850s the distillery was managed by Mrs Margaret Sutherland.

In 1867 the tenancy passed to Andrew Mackenzie (aged 24), the son of Matheson's estate factor and his brother. He was tasked with 'expanding the business' and, helped by his younger brother, Charlie, he set about doing just that, doubling capacity by 1874 with a new still-house, equipped with unusual stills (see below). Their descendants managed (and after 1891 owned) Dalmore until 1960, when the company amalgamated with Whyte & Mackay.

The brothers Charles, Alexander and Andrew Mackenzie took over in 1867; their descendants ran Dalmore until 1960, when they amalgamated with Whyte & Mackay, a company with which they had friendly relations from the outset.

Ownership of Dalmore Distillery passed to the United Spirits Division of the United Breweries Group of India in 2007, when that company acquired Whyte & Mackay. Plans were laid to triple capacity, and work had begun when it was announced (in 2012) that United Spirits was to be sold to Diageo. The deal was completed in 2014, but Diageo was obliged by the Office of Fair Trading to immediately sell Whyte & Mackay. The company, with its distilleries and brands, was bought by Emperador, the Philippine brandy distiller, in 2014 (see *Invergordon*).

CURIOSITIES: Andrew Mackenzie was an innovator. He was among the earliest distillers to finish his

Dalmore 'Trinitas' 64YO was the first bottle to sell for £100,000, on 14 October 2010. It was so named because only three bottles were filled. One was drunk at WhiskyLive London 2010; the other is available from Harrods for £120,000.

whisky in ex-sherry casks after maturation for five or six years in 'distillery wood' (either refill or new oak casks). Through Alexander Matheson's contacts in Asia and Australia, he was among the first Scotch whisky distillers to sell single malt in the Far East – and the very first to introduce it to Australia, which soon became the largest export market for Scotch whisky (remaining so until 1938).

Dalmore has several unusual or unique features. The large wash charger – around six metres across – is made of pine. Formerly, it was equipped with two paddles connected to a long wooden tiller; the operator would walk round the vessel, pushing the tiller and driving the paddles, in order to agitate the wash and prevent sediment entering the wash stills before the wash itself.

It is claimed that the stills are the oldest in the Highlands. Part of one of them dates from 1874. Four (the wash stills) have flat tops, rather than the usual 'swan necks'. This makes for a heavier, more characterful spirit. The other four (the spirit stills) have unique water-jackets around their necks (first fitted in 1839), so the copper is continually cooled, increasing reflux and making for a lighter spirit. Curiously, the condensers for the spirit stills are mounted horizontally, outside the stillhouse.

Furthermore, one of the spirit stills is twice as big as the other three – another unique feature. The spirit coming from this still has a different character to that from the others – citric fruits and aromatic spices, while the smaller stills create rich and robust musky flavours.

There are four spirit safes in the stillhouse, one of them of unusual design and considerable antiquity.

The 12-point (or 'royal') stag which embellishes bottles of Dalmore was adopted in 1886 and recollects the fact that this emblem was granted to an ancestor of the Mathesons and Mackenzies,

Dalmore was the first single malt whisky exported to Australia (in 1870).

who rescued King Alexander III from being gored by a stag in 1263.

The name means 'the Big Meadow', and the situation, on the alluvial plain overlooking the Cromarty Firth, supports this. During World War I the distillery warehouses were requisitioned by the Admiralty for the manufacture of mines. The pier below the distillery was built by them and is known as the Yankee Pier.

Drew Sinclair, who retired as manager in 2006 (and died soon after) worked at Dalmore for 40 years.

One of a dozen bottles of Dalmore 62YO sold privately in April 2005 for £32,000, then a world record. (In truth the bottle contained whisky from 1868, 1878, 1926 and 1939.) The purchaser opened and drank the bottle immediately, with friends.

RAW MATERIALS: Floor maltings until 1956, then converted to Saladin boxes until 1982. Now unpeated malt from Bairds, Inverness. Process water from Loch Kildermorie, on the slopes of Ben Wyvis; cooling water from River Averon or Alness.

PLANT: Semi-Lauter mash tun (9.2 tonnes); eight Oregon pine washbacks. Four lamp-glass, flat-top wash stills (three of 13,411 litres, one of 30,000 litres); four boil-pot spirit stills, fitted with water jackets to cool the necks of the stills (three of 8,865 litres, one of 19,548 litres). All indirect fired by steam. Shell-and-tube condensers.

MATURATION: Mix of ex-sherry, ex-bourbon and refill hogsheads. Matured on site mainly in dunnage warehouses and in Leith.

STYLE: Heavy, oily and musky (from the smaller stills); lighter and more citric (from the larger still).

MATURE CHARACTER: The overall style of Dalmore new-make is appropriate for sherry-wood maturation.

The nose is rich and sherried, with sweet malt, fruit cake, orange peel and marzipan. Medium- to full-bodied, the texture is mouth-filling and the taste sweet rather than dry. Long finish.

Da

REGION
Speyside
ADDRESS
Carron, Banffshire
OWNER
Chivas Brothers
VISITORS
No
CAPACITY
10m L.P.A.

Dalmunach

HISTORICAL NOTES: Pernod Ricard announced in October 2012 that Imperial Distillery closed in 1998 and put up for sale as residential flats in 2005 (see entry p.242) would be demolished and rebuilt. Work on the new distillery began in 2013 and by the end of that year the site was clear, except for one old warehouse.

The new distillery cost £25 million. It is strikingly modern in design and operation, but sits well in the wooded site on the bank of the River Spey. It won the Royal Institute of British Architects Award for Scotland in 2015. Ingvar Ronde, publisher of the invaluable *Malt Whisky Yearbook 2016*, describes it as 'not only one of the largest, but one of the most beautiful [distilleries] in Scotland'. It takes its name from a nearby pool in the Spey and was opened in June 2015 by Nicola Sturgeon, Scotland's First Minister.

CURIOSITIES: It was speculated – but only briefly – that the new distillery might be called 'Imperial II', just as Chivas Brothers' new Glenburgie distillery retained the name of the old one. However, the term 'imperial' is now deeply unfashionable, and since Chivas owns a blended Scotch brand of the same name, it would also contravene the Scotch Whisky Regulations 2009.

The distillery's design was apparently inspired by the shape of a sheaf of barley. The interior is spacious and light, with glass walls at each end of the stillhouse. The still shapes replicate those of the old distillery, and the stills themselves are arranged in a circle around the spirit safe, rather than in a line. Other original features from the previous distillery have been incorporated to give a sense of heritage: red brick from the original mill building has been reclaimed to create a feature wall in the new entrance area, and Oregon pine from the original washbacks has been used to form an entrance 'pod' and adorn the gable walls of the new tun room.

Dalmunach's construction was managed from start to finish by Douglas Cruickshank, who retired as Chivas Brothers' Production Director in 2013 to concentrate on the new build – a fitting end to Douglas's long career, which began at the age of fifteen at Imperial Distillery.

RAW MATERIALS: Unpeated malt from independent maltsters; water from the original springs on site as well as taking cooling water from the River Spey.

PLANT: Stainless steel full-Lauter tun (12 tonnes). Sixteen stainless steel washbacks. Four lamp-glass wash stills – the same as original Imperial shape (each 30,000 litres charge), four plain spirit stills (20,000 litres charge). All indirectly fired and equipped with shell-tube condensers.

MATURATION: Predominantly first-fill and refill U.S. oak hogsheads, with some Spanish oak butts.

STYLE: Sweet with notes of creamy toffee, fruit and esters; Speyside style. The whisky will be used in Chivas Brothers blends, predominantly Chivas Regal, Royal Salute and Ballantine's.

Dalwhinnie

REGION
Highland (Central)
ADDRESS
Dalwhinnie,
Inverness-shire
PHONE
01540 672219
WEBSITE
www.malts.com
OWNER
Diageo plc
VISITORS
Visitor centre and shop
CAPACITY
▮▮▮ ▮▮▮▮

HISTORICAL NOTES: Dalwhinnie is 'the meeting place', the place where drove roads from the north and west met those coming out of Strathspey and headed south; in the 1730s these were replaced by military roads, under the supervision of General Wade, who also built a straggling village here.

The distillery itself was originally named Strathspey. It was built in 1897 by three men from the neighbouring town of Kingussie, but they soon ran into financial difficulties and sold up. The new owners changed the name to Dalwhinnie and employed Charles Doig to make improvements, then sold in 1905 to Cook & Bernheimer, the largest distillers in America at that time. This was the first time a Scottish distillery came under foreign ownership, but it was to last only 14 years, when Prohibition was announced in the U.S.A. Dalwhinnie was sold again, to the well-known blending firm Macdonald, Greenlees & Williams. The latter joined D.C.L. in 1926, and Dalwhinnie was licensed to James Buchanan & Company.

The distillery was badly damaged by a fire in 1934 – until that year there was no electricity in the village and the distillery was lit by paraffin lamps. It was rebuilt and opened again in April 1938, the original stillhouse becoming the present day tunroom, only to be closed again during World War II.

There was another refit in the 1960s. The stills were converted to indirect heating by steam in 1961 (originally a coal boiler, converted to oil 1972); malting on site ceased in 1968; British Rail closed the private siding off the main line, which runs at the back of the distillery, in 1979. Between 1992 and 1995 the distillery was closed for a major refurbishment.

Dalwhinnie single malt became well-known when it was selected by U.D. to represent the Highland style in the Classic Malts series.

The distillery's priorities are 'Family, Friends and Flavour'.

CURIOSITIES: Until Braes of Glenlivet (now Braeval) was built in 1973, Dalwhinnie was the highest distillery in Scotland, at 1,073 feet above sea level. It is a remote spot (in spite of the nearness of the main north trunk road, the A9), and the coldest distillery, with an average annual temperature of 6°C. Since 1973 Dalwhinnie has operated as a 'Climatological Station'. Every day, including Christmas and New Year, readings of temperature, wind strength, humidity, visibility, frost and hours of sunshine must be recorded and sent to Edinburgh. This job used to be the (unpaid) responsibility of the manager, but is now shared by the operators.

Hamish Christie, stillman, recalls playing football in 1963 on a pitch covered with five feet of hard-packed snow, using as goalposts the single visible foot of the six-foot-tall poles which marked the roadway.

'Its wind-blasted location made a visit, even in May, seem like a major expedition to a forgotten outpost', wrote Philip Morrice in 1987. It is not uncommon for Dalwhinnie to be snowbound for weeks – but this may change with global warming.

In 1986 the worm tubs here were replaced by shell-and-tube condensers (from Banff Distillery), but this changed the spirit character so much that the move was reversed and the worms reinstated, in 1995.

In such a remote spot, it is common for families to work here generation after generation. Maureen Stronach, Head Guide, who has worked at the distillery for 35 years, succeeded her father and grandfather (both of whom were brewers) and three uncles. Her late husband was a stillman, and now her brother Hamish (with 21 years' service) and son are both stillmen/operators at Dalwhinnie.

RAW MATERIALS: Floor maltings until 1968; now lightly peated malt from Roseisle. Soft water from Lochan an Doire-uaine (2,000 feet) via Allt an t'Sluie Burn.

PLANT: Full-Lauter mash tun (7.3 tonnes). One Oregon pine washback and five Siberian larch washbacks. One plain wash still (17,500 litres charge); one plain spirit still (16,200 litres charge). Direct fired until 1961, now indirect. Worm tubs. The lye pipe from the spirit still is sprayed with cold water before it enters the worm tub, to encourage early condensation, and thus lack of copper uptake.

MATURATION: Mainly ex-bourbon refill casks, matured in the Central Belt.

STYLE: Full-bodied, sweet, heather-honey.

MATURE CHARACTER: Dalwhinnie is a remarkably viscous malt, with a big mouthfeel. The nose is sweet, with heather pollen, honeycomb, and moorland scents. The taste is soft and smooth, starting sweet and drying out with a whiff of peat smoke. Medium-bodied.

Deanston

De

REGION
Highland (South)

ADDRESS
Doune, Perthshire

PHONE
01786 843010

WEBSITE
www.deanstonmalt.com

OWNER
Distell Group Ltd

VISITORS
By appointment

CAPACITY
3m L.P.A.

HISTORICAL NOTES: Deanston Distillery was created within an historic cotton mill, designed in 1785 by Richard Arkwright (pioneer of steam-powered spinning). It is the classic example of a mill-distillery, of which there are several in Scotland, for the simple reason that both require copious supplies of fast-running pure water. It stands on the edge of the River Teith, near the picturesque Doune Castle, and operated until 1965.

The distillery conversion was the brain-child of Brodie Hepburn Ltd (whisky brokers in Glasgow, owners since 1953 of Tullibardine Distillery and builders of Macduff Distillery in 1963). They took a 30% share, in partnership with the owners of the mill, James Findlay & Company. A water turbine and stand-by generator were already in place, but four solid floors had to be removed to make way for two pairs of stills. Production commenced in October 1969, and the first Deanston Single Malt was released in 1974.

The original plan was to link production to the development of a major brand of blended whisky, (Old Bannockburn), but this never happened and in 1972 Invergordon Distillers took control. The distillery was silent from 1982 to 1990, when it was sold to Burn Stewart Distillers for £2.1 million in cash and brought back into production. Burn Stewart was acquired in 2002 by C.L. World Brands (based in Trinidad), and C.L. World Brands was acquired by the Distell Group of South Africa.

CURIOSITIES: 'Dean' and 'Doune' come from the Gaelic 'dun', a hill-fort. Doune Castle is 'one of the largest, best-preserved and best-restored examples of late fourteenth-century military-domestic architecture in Scotland' (*The Blue Guide to Scotland*).

A gin distillery was commissioned within Deanston in 1995; still in place and capable of being brought back into service if required.

Doune Castle was used in *Monty Python and the Holy Grail* (1974) as Camelot, Castle Anthrax and Swamp Castle!

Brodie Hepburn claimed in an advertisement in the *Scottish Licensed Trade News* (1966) to be 'the oldest established whisky broker'. The company was bought by Invergordon Distillers in 1971.

Burn Stewart dates back to the 1940s, but it was a management buy-out led by Bill Thornton in 1988 that created the present company. They paid £7 million; the company went public three years later, with a capitalisation of £83 million, which allowed the Directors to build offices and bottling facilities at East Kilbride in 1992/93 (expanded 1995) and buy Tobermory Distillery in 1993. Bunnahabhain Distillery was acquired in 2003 (along with the successful Black Bottle blend). The company's blends are prepared at Airdrie, which also has extensive warehousing.

The interior of Deanston Distillery features in Ken Loach's film *The Angels' Share*.

RAW MATERIALS: Unpeated malt from independent maltsters. Soft water from the River Teith. Process water from stream in the Trossachs; cooling water from the River Teith.

PLANT: Cast iron mash tun (11.2 tonnes). Eight Corten steel washbacks. Two boil-ball wash stills (8,500 litres charge); two boil-ball spirit stills (6,500 litres charge). Indirect fired. Shell-and-tube condensers.

MATURATION: 40,000 casks on site in a former weaving shed known as Adelphi Mill; remainder at Airdrie.

STYLE: Waxy, light, fruity.

MATURE CHARACTER: Lightly oily on the nose, with cereal notes. The taste is malty and fruity and lightly nutty, starting sweetish and finishing dryish. Light- to medium-bodied.

Dufftown

REGION
Speyside

ADDRESS
Dufftown, Moray

PHONE
01340 822100

WEBSITE
www.thesingleton.com

OWNER
Diageo plc

VISITORS
No

CAPACITY
6m L.P.A.

HISTORICAL NOTES: The distillery was converted from a meal mill in 1895/96. It is located just outside Dufftown in the Dullan Glen, and the site was chosen by Peter Mackenzie and Richard Stackpole, of P. Mackenzie & Company, wine and spirits brokers in Liverpool (owners of The Real Mackenzie blend). Peter Mackenzie was born in Glenlivet, and his firm had bought Blair Athol Distillery in 1882. The founders included a local farmer and owner of the original mill, John Symon (who also supplied barley from his Pittyvaich Farm), and John Macpherson, a local solicitor. A year after its foundation, Peter Mackenzie formed a limited company to acquire full ownership of Dufftown-Glenlivet and the other assets of Mackenzie & Company.

In 1933 Mackenzie & Company (Distillers) Ltd was bought by Arthur Bell & Sons, and Dufftown-Glenlivet became a key filling for Bell's Extra Special. The distillery was expanded from two to four stills in 1968, to six in 1974 and to eight in 1979 (later reduced to six again).

Bell's was acquired by Guinness plc in 1985; Guinness went on to take over D.C.L. in 1987, so Dufftown now falls within Diageo's estate.

Dufftown is Diageo's third largest malt whisky distillery, with a capacity of six million litres of pure alcohol a year. Since 2007, the make has been bottled as The Singleton of Dufftown for the European market (see also *Glendullan* and *Glen Ord*).

CURIOSITIES: Dufftown-Glenlivet was the sixth distillery to be built at Dufftown.

Unusually, the spirit stills are larger than the wash stills.

Dufftown village itself was founded in 1817 by James Duff, fourth Earl of Fife, as a means of finding employment for men returning from the Napoleonic Wars. He had distinguished himself

The men of Mortlach (Dufftown's original distillery) resented the arrival of a newcomer and sought to divert Dufftown's water supply on more than one occasion.

during the Peninsular War, achieving the rank of Major General, and was M.P. for Banffshire 1818–27. The town was originally named Balvenie after the large medieval castle that stands on its edge, but soon changed its name to that of its founder. (See *Balvenie*.)

RAW MATERIALS: Floor maltings until 1968, now unpeated malt from Burghead. Soft process water from Jock's Well in the Conval Hills (supplemented by the Convalley Springs), cooling water from the River Dullan.

PLANT: Full-Lauter mash tun (11 tonnes). Twelve stainless steel washbacks. Three plain wash stills (15,000 litres charge), three plain spirit stills (15,000 litres charge). All indirect fired, and run hot. Shell-and-tube condensers, with after-coolers.

MATURATION: Ex-bourbon refills, with a small amount in ex-sherry refills.

STYLE: Malty, nutty.

MATURE CHARACTER: The nose is Speyside-sweet, fruity and cereal-like (bruised apples, pears), with some butterscotch. The taste is predominantly sweet, with cereal notes and light toffee. Medium-bodied.

Dumbarton Grain Distillery (Demolished)

ADDRESS
2 Glasgow
Road, Dumbarton
Dunbartonshire
OWNER
Allied Distillers
CAPACITY
25m L.P.A.
CLOSED
2002

HISTORICAL NOTES: Dumbarton Distillery was built in the town of the same name by Hiram Walker (Scotland) Ltd in 1938 from millions of red bricks, an unusual material in Scotland. The site was formerly MacMillan's shipyard, and the distillery was designed in North America, the continuous stills coming from the Vulcan Copper & Supply Company of Cincinnati. In its day it was the largest grain distillery in Scotland.

Inverleven (malt whisky) Distillery was built within the Dumbarton complex at the same time as the grain distillery, and in 1959 a Lomond still was added (see Miltonduff). In 1965 the first dark-grains plant in the U.K. was installed (based on the American system).

Hiram Walker was bought by Allied Lyons in 1987, and Dumbarton Distillery was closed in 2002. It has now been demolished.

CURIOSITIES: Water came from Loch Lomond, which gave its name to the Lomond still.

Dumbarton used only maize, shipped mainly from the U.S.A. and France. This produces a robust and oily spirit. The stills were not capable of processing other grains, and it would seem that their original design did not incorporate enough copper (a purifier) to remove heavier compounds in the make.

Hiram Walker Gooderham & Worts Ltd was the largest distiller in Canada (Canadian Club was its leading brand). Keen to enter the Scotch whisky market, the company bought Ballantine's in 1935, together with extensive stocks of mature whisky but no distilleries. It was imperative to acquire production facilities, so Glenburgie and Miltonduff Distilleries were acquired in 1936 (see entries), and then Dumbarton was built.

Edradour

REGION
Highland (South)

ADDRESS
Milton of Edradour, near
Pitlochry, Perthshire

PHONE
01796 472095

WEBSITE
www.edradour.com

OWNER
Signatory Vintage Scotch
Whisky Co. Ltd.

VISITORS
Visitor centre, shop and
tasting bar

CAPACITY
120,000 L.i.A.

*Until recently
Edradour was
the smallest
Highland distillery,
producing just
twelve casks
of spirit a week.
It is also one
of the most
popular distillery
destinations
in Scotland,
welcoming around
55,000 visitors
a year.*

HISTORICAL NOTES: Edradour is a classic farm distillery, was the smallest in Scotland until 2005, and remains among the prettiest. It stands as a reminder of how many distilleries will have looked, and operated, during the 19th century.

In all likelihood it was established by one Duncan Forbes in 1825 (named Glenforres), but the foundation on the present site was in 1837, when a group of farmers leased a strip of land beside the Edradour Burn from the Duke of Atholl. Their leader was Mungo Stewart; Duncan Forbes and six other farmers joined the co-operative.

After 1841 they were led by John McGlashan, who continued to manage the distillery until 1877, although after 1853 the license was in the name of another local farmer, James Reid.

In 1885 the distillery was transferred to John Mackintosh, son of one of the original founders. He increased production and transformed Edradour into a successful commercial enterprise. When he died in 1907 he left the distillery to his nephew Peter, whose father (John's brother) was the excise officer at Ord Distillery.

These were difficult times for the whisky industry, and Peter was further hindered by bad health. Between 1920 and 1932 cask sales dropped from 90 to 21. In early 1933 he sold the distillery and two cottages to William Whiteley for £1,050.

Whiteley's subsidiary, J.G. Turney & Sons, was a customer of Edradour, and used the make in his blends House of Lords and King's Ransom (the latter introduced in 1928 and reputed to be the most expensive whisky of its day).

He retired in 1938 and sold the company and distillery to his American distributor, Irving Haim, who left Edradour as it was, except for installing electricity in 1947. On his death in 1976, the distillery was briefly owned by a consortium, then bought by Campbell Distillers (a subsidiary of Pernod Ricard) in 1982. Four years later Edradour was bottled as a single malt for the first time, at 10YO, and the former malt barn was developed into a visitor centre and shop.

In July 2002, Edradour was bought for £5.4 million (£3 million of which was maturing stock) by Andrew Symington, owner of Signatory Vintage Malt Whisky Company. Many improvements have been made, on site, including building a bottling hall (2007) and a large new warehouse (2010), incorporating 'Caledonia Hall', a function suite used for whisky-themed events. The visitor centre within the old maltings was refurbished in 2011 – Edradour is one of the most visited distilleries in Scotland, welcoming around 55,000 people a year.

Mr Symington is currently planning to build a new distillery on an adjacent site, to be named 'Edradour II'. It was to be named Ballechin, after a Perthshire distillery which closed in 1927, but this was disallowed since he uses the name for a peated expression of Edradour.

CURIOSITIES: William Whitely (1861–1942) was a tough Yorkshireman and a canny marketer. He had started as a salesman for a wine and spirits company, but was dismissed in 1908 for acting beyond his remit. In 1914 he purchased J.G. Turney & Son, another wine and spirits merchant, based in Leith, but selling mainly in export markets. With the introduction of Prohibition in the U.S., Whiteley appointed Frank Costello as his 'U.S. Sales Consultant' with an annual salary of $5,000. It may have been Costello who described Whiteley as 'The Dean of Distillers', and the name stuck.

Costello was a leading figure in the Mafia, becoming its Capo di tutti Capi (and the model for 'The Godfather' in Mario Puzo's books). Described as a 'visionary gangster', he was a major bootlegger and controlled numerous speakeasies and clubs in New York. Whiteley's business was handed by Costello's side-kick,

Irving Haim, and when Whitely retired in 1938, he acquired J.G. Turney & Son.

RAW MATERIALS: Unpeated malt from Bairds, Edinburgh; peated malt from Inverness. Pre-1975 the malt was ground on site by mill-stones; between 2002 and 2007 it arrived ready milled. Soft process water from a spring on Moulin Moor, cooling water from Edradour Burn.

PLANT: Cast iron, rake and plough mash tun made in 1910 (1.1 tonnes). Wort cooled by a Morton Refrigerator (the last left in the whisky industry; made in 1933 and installed in 1934). Two Douglas fir washbacks. One plain wash still (4,200 litres charge); one boil ball spirit still (2,000 litres charge), equipped with a purifier. Both indirect fired. Worm tubs on both stills.

MATURATION: Edradour is matured in a combination of refill U.S. oak hogsheads and European oak butts; around a dozen different kinds of first-fill wine casks are also used, either for maturing or finishing. Ballechin is primarily matured in first-fill ex-bourbon barrels, with some first-fill sherry and refill sherry. In both makes, no cask is used more than twice.

STYLE: Edradour fruity, pear-like, with a good body. Ballechin peaty.

MATURE CHARACTER: Clean and fresh, with floral as well as fruity notes – almond blossom and dog roses, with cider apples and citric notes. Good texture, with a sweet start, slightly mouth-cooling (minty), drying in the medium-length and lightly spicy finish.

Ed

Eden Mill

REGION
Lowland
ADDRESS
Guardbridge, by St
Andrews, Fife
PHONE
0800 086 8290 and
01334 834038
WEBSITE
www.edenmill.com
OWNER
St Andrews Brewers Ltd.
VISITORS
Shop and tours (whisky,
gin and beer), seven days
10 a.m. to 5 p.m. (best
to book).
CAPACITY
100,000 L.P.A.

HISTORICAL NOTES: Eden Mill is on the site of Seggie Distillery, founded by William Haig in 1810 and later taken over by his brother, John, who acquired Cameronbridge Distillery in 1824. Some of the original buildings still stand. Whisky from Seggie is recorded in 1857 as a component of Arthur Bell's Perth blends. The distillery operated until at least 1865 (when John Haig formed the Scotch Distillers Association, which became the Distillers Company Limited), and in 1872 was converted into a large paper mill. This operated until 2008; the site was then bought by the University of St Andrews.

Part of the old mill was leased by a consortium led by Paul Miller, former sales director of Molson Coors and marketing director at Glenmorangie, who opened a craft brewery in 2012 and expanded into whisky and gin production in November 2014. In 2015, 100,000 L.P.A. of malt spirit was made, and the owners have no plans to increase capacity: 'We do not mass produce. We fill only a handful of hogsheads each week. We're big on keeping it small!' As well as whisky, Eden Mill produces an exotic range of craft gins.

Work is underway for the current distillery to relocate eighty yards north into an original distillery building in spring 2017.

CURIOSITIES: With five degree-qualified distillers, each producing one hogshead per week, Eden Mill believes it has more qualified distillers per litre of whisky than any other in Scotland!

Eden Mill also spends a significant time on developing blends and on small cask maturation work. A limited-run release of their first blend with differentiated finishing in quarter casks formerly Pedro Ximénez, American virgin oak, Oloroso and bourbon will be released in 2016. Two Eden Mill distillers are also trained coopers.

RAW MATERIALS: Locally grown Golden Promise barley is the distillery's exclusive 'pale malt', but it also takes advantage of its small batch production runs by utilising some other malt types and has three core spirits: a 100% pale malt; 90% pale and 10% chocolate malt; and a 90% pale and 10% crystal and brown malt. From 2017 it will be using water from the Guardbridge Mill reservoir which was used by Seggie Distillery.

PLANT: Stainless steel semi-Lauter tun (850 kg). Five stainless steel washbacks, two of which are used for whisky. Two Hoga wash stills (each 1,000 litres charge), one Hoga spirit still (1,000 litres charge). All with shell-tube condensers; all stills indirectly fired. All distillations are pot still batch distillations.

MATURATION: Oloroso and Pedro Ximénez octaves, barrels and hogsheads for the core style, however some virgin American and European oak casks are used for anticipated special releases. Over 80% of Eden Mill spirit is matured in virgin or first-fill wood.

STYLE: Sherry matured and robust for a Lowland malt with a slight coastal hint.

Fettercairn

REGION
Highland (East)

ADDRESS
Fettercairn,
Laurencekirk, Angus

PHONE
01561 340244

WEBSITE
www.fettercairn
distillery.co.uk

OWNER
Whyte & Mackay Ltd

VISITORS
Visitor centre

CAPACITY
2.2m L.P.A.

*Some single cask
bottlings on site are
for sale exclusively
through the visitor
centre.*

HISTORICAL NOTES: Fettercairn is in the heart of the fertile Mearns district, celebrated by the writer Lewis Grassic Gibbon (*Sunset Song* etc.). The distillery was established by the laird of Fasque, Sir Alexander Ramsay, in 1824. He sold the estate to Sir John Gladstone, father of Prime Minister William Ewart Gladstone, in 1830. Although managed by tenants, the Gladstone involvement remained until 1923, when the distillery was sold to Ross & Coulter, whisky brokers since 1919 (and later owners of Bladnoch and Bruichladdich Distilleries). They mothballed it from 1926 to 1939, then sold it to a subsidiary of National Distillers of America (which would ultimately own Ben Nevis, Bruichladdich, Lochside, Glenesk and Glenury Royal Distilleries). Extended from two to four stills in 1966/67.

Fettercairn was bought by Tomintoul-Glenlivet Distillery Company in 1971, and the latter was acquired by the Scottish & Universal Investment Trust, owner of Whyte & Mackay Distillers, two years later. They built an extensive effluent disposal plant adjacent to the distillery in 1980. In 2007 Whyte & Mackay was acquired by the Indian brewer and distiller U.B. Group. The spirits division of U.B.G. was bought by Diageo in 2012, who sold it on to the Philippine distilling company Emperador in 2014 (see *Invergordon*).

CURIOSITIES: The water-cooling system on the spirit stills is unique. Cold water cascades down the necks of the stills from a collar, collecting in a trough above the shoulder to be drawn off and re-used in the boiler. A further curiosity is the fact that each pair of stills has its own spirit safe – referred to as No. 1 Side and No. 2 Side – although the makes of each are mixed.

The product of the distillery was sold as Old Fettercairn until 2002, when the core expression was rebranded and repackaged as Fettercairn 1824.

RAW MATERIALS: Floor maltings until 1960; now unpeated malt from independent maltsters and, since 2005, heavily peated (55ppm phenols) for a couple of weeks a year for blending purposes. Process water from springs in the Grampian Hills which rise behind the town; cooling water from the Caulecotts Burn.

PLANT: Cast iron traditional mash tun (4.88 tonnes); eight Oregon pine washbacks.

Two plain wash stills (13,000 litres charge); two plain spirit stills (13,500 litres and 11,500 litres charge), both with water jackets. All indirect fired. Shell-and-tube condensers (all stainless steel until 1995).

MATURATION: Mixed ca-sherry, ex-bourbon and refill hogsheads. Matured mainly at Invergordon; some in dunnage warehouses on site, where there is capacity for 30,000 casks.

STYLE: Butterscotch, walnuts and spice.

MATURE CHARACTER: The nose is sweet and malty, with traces of damp wool and mixed nuts. The taste is sweetish to start, drying with nuts, biscuits and a whiff of smoke. Slightly oily. Medium-bodied.

Girvan Grain Distillery

ADDRESS
Girvan, South Ayrshire
PHONE
01465 713091
OWNER
William Grant & Sons Ltd
CAPACITY
105m L.P.A.

HISTORICAL NOTES: The post-war boom in demand for Scotch, combined with grain rationing – the industry was not completely deregulated until 1953 – and an acute shortage of aged whiskies (until 1959), meant that the larger companies (notably D.C.L.) limited the amount of both malt and grain whisky available to smaller producers.

During the late 1950s/early 1960s several companies built new grain whisky distilleries to meet the demand: Invergordon (Invergordon Distillers Ltd, 1959), Strathmore (North of Scotland Distilling Company, 1960) and Moffat/Garnheath (Inver House Distillers, 1965). William Grant & Sons, owners of Glenfiddich and Balvenie Distilleries, and of the successful blend Standfast, built Girvan in 1963/64. Within a year they had added a malt whisky distillery to the site, named Ladyburn (see entry p.264).

Key reasons for choosing the small Ayrshire port of Girvan were the water supply, the availability of labour and access to both the sea and Lowland blending houses.

CURIOSITIES: In 1962 the ever enterprising Grant's of Glenfiddich proposed running an advertising campaign for their popular Standfast blend on the nascent commercial TV channel in the U.K. The mighty D.C.L. responded by threatening to cut off their supply of grain whisky, and it is said that this influenced the family's decision to build their own grain whisky distillery. The driving force behind the project was Charles Grant Gordon, who died in 2014.

The dark-grains plant which was also built on the site has the largest filter presses in the world for separating solids and yeast debris.

During World War II the site was used by I.C.I. to make munitions, supplied by two 1m-gallon water tanks in the hills behind.

Until 1986 the grain used was maize, brought in

The distillery has an austere appearance, influenced by the Bauhaus design style.

from the U.S.A. by sea. The popular Hendrick's gin is also made at Girvan in a small separate distillery, and since 2007 Ailsa Bay (malt) Distillery has also been operating on the site.

For many years Girvan was available in the U.S.A. and in duty free under the Black Barrel label; in 2013 and 2014 this was replaced by Girvan No. 4 Apps, Girvan 25YO and Girvan 30YO Patent Still Grain whiskies.

RAW MATERIALS: The cereal base is home-grown wheat and malted barley. Water from Penwhapple Loch.

PLANT: Until 1995 employed a continuous mashing plant which integrated cooking and mashing. Production started in 1964 with three stills, called 'apparatus's' or 'Apps'. No. 1 App is a Coffey still, made entirely of copper, used solely to produce grain whisky, but only occasionally used today. No. 2 App was a 'clean-up column' to make neutral spirit from No. 3 App; No. 3 App was also a Coffey still, but could produce both whisky and neutral spirits. The last two are no longer used. No. 4 and No. 5 Apps were made by the Finnish state distiller Alco, and installed in 1992. They employ patented multi-pressure distillation systems, with a unique vacuum distillation process in the analyser (the wash column), heated by vapour off the rectifier.

MATURATION: Mainly first-fill and refill American oak ex-bourbon barrels. 1.62 million casks on site; 1.5 million L.P.A. sent to Wm Grant's Glenfiddich site at Dufftown.

STYLE: The spirit from No. 1 App is much heavier and oilier than that from No. 4 and No. 5 Apps. The overall style is clean, estery and sweet.

MATURE CHARACTER: Much influenced by the wood – fresh and fruity (apples), with vanilla sponge; sweet with light acidity and some spice; coconut and light oak. Older expressions are denser and more spicy.

G1 Glasgow

REGION
Lowland

ADDRESS
Hillington Business Park

WEBSITE
www.glasgowdistillery.com

OWNER
The Glasgow Distillery Company Ltd

VISITORS
No

CAPACITY
250,000 L.P.A.

HISTORICAL NOTES: With backing from Asian investors, Liam Hughes and Ian McDougall began making gin in October 2014 and whisky on March 20th 2015. Current production is around 120,000 LPA, but can be increased to 250,000. The distillery manager is Jack Mayo, and the 'consulting distiller' is David Robertson, former manager of The Macallan Distillery, who is currently seeking permission to establish a new distillery in Edinburgh.

Glasgow Distillery also makes Makar's Gin and currently offers a single Speyside malt named Prometheus.

CURIOSITIES: The distillery has a laboratory attached to experiment with different raw materials, varying styles of new-make spirit and different casks, all in the interest of new product development, especially in relation to using a gin still named 'Annie' to create a range of spirit expressions.

RAW MATERIALS: Unpeated malt from independent maltsters (currently Muntons). Water from Loch Katrine.

PLANT: Stainless steel semi-Lauter mash tun (one tonne). Four stainless steel washbacks. One boil-ball wash still (2,400 litres charge), one boil-ball spirit still (1,500 litres charge), both indirect fired. Shell-and-tube condensers. Also a rectifying still named 'Annie' for making gin.

MATURATION: On site. Many different cask varieties and sizes will be used, with different finishes, maturation periods and conditions.

STYLE: 'Metropolitan' style, similar to Speyside. Stone fruit flavours and creamy texture.

Glen Albyn Demolished

REGION
Highland (North)

ADDRESS
by Muirtown Basin,
Merkinch, Inverness

LAST OWNER
D.C.L./S.M.D.

CLOSED
1983

Glen Albyn had a reputation as being 'the classic example of Highland style'.

HISTORICAL NOTES: The distillery was built on the ruins of Muirtown Brewery, beside the Caledonian Canal, in 1840 by James Sutherland, Provost of Inverness. The site provided easy access to southern markets by sea, but the distillery was badly damaged by fire nine years after it opened, and in 1855 Sutherland was sequestrated. Briefly used as a flour mill, the buildings lay unoccupied for 20 years, before being bought (in 1884) by William Grigor, a grain merchant, who built a new and larger distillery on the site.

Manager John Birnie, in partnership with Charles Mackinlay & Company, bought the site across the road and built Glen Mhor Distillery in 1892 (see entry). Mackinlay & Birnie bought Glen Albyn in 1920, and the company was acquired by D.C.L. in 1972 and transferred to S.M.D. Both distilleries were closed in 1983, and the Glen Albyn site is now a supermarket.

CURIOSITIES: During World War I, Glen Albyn became a U.S. naval base for the manufacture of mines, according to one source, and an Admiralty depot making boom defences according to another.

The distillery's floor maltings were replaced by Saladin boxes in 1954.

Although only a matter of yards from each other, the make from Glen Albyn was entirely different from that of Glen Mhor, and rated more highly.

G1   Glenallachie

REGION
Speyside

ADDRESS
Aberlour, Moray

PHONE
01340 871315

OWNER
Chivas Brothers

VISITORS
No

CAPACITY
4m L.P.A.

HISTORICAL NOTES: The 'Glen of the Rocky Place' (*Gleann Aleachaidh*) runs up behind Aberlour towards Edinvillie. It is not very rocky, although the mass of Ben Rinnes rises close beside the distillery, which draws its production water from springs on the mountainside.

It was commissioned by Mackinlay-Macpherson (then a subsidiary of Scottish & Newcastle Breweries) in 1967, primarily to contribute Speyside fillings to the Mackinlay blend (which at that time was among the top five bestsellers in the U.K.), and designed by William Delmé-Evans. The architect was Lothian Barclay, son of the whisky entrepreneur, James Barclay (see *Strathisla*). The design is modern and efficient, and very 1960s.

Mackinlay's was sold to Invergordon Distillers in 1985, and Glenallachie was mothballed, then closed. In 1989 it was sold to Campbell Distillers (Pernod Ricard), and went back into production. Chivas Brothers became Pernod Ricard's distilling division in 2001. Glenallachie was expanded from three to four million L.P.A. capacity in 2012/13.

CURIOSITIES: William Delmé-Evans has been compared to Charles Doig, the greatest distillery architect of the late nineteenth century. Having completed Tullibardine and Jura Distilleries, he turned to Glenallachie, his last project.

He was such a stickler for efficiency that he even marked the light-bulbs in the distillery with the date on which they were installed, to monitor how long they lasted!

RAW MATERIALS: Unpeated malt from independent maltsters. Water from springs on Ben Rinnes via two dammed streams flowing into the Lour Burn. Cooling water is collected in an attractive pond by the distillery.

PLANT: Semi-Lauter mash tun (9.2 tonnes); six stainless steel-lined washbacks.

Glenallachie comes from Gleann Aleachaidh *– the glen of the rocky place.*

Two lamp-glass wash stills (23,000 litres charge); two plain spirit stills (16,000 litres charge). All designed by Delmé-Evans. Indirect fired. Horizontal shell-and-tube condensers.

MATURATION: Mainly refill U.S. hogsheads, some first-fill. Twelve racked and two palletised warehouses on site, the rest matured at other Chivas sites.

STYLE: Sweet, grassy and estery.

MATURE CHARACTER: Speyside sweetness, with fruity-floral notes. Some detect a whiff of smoke. Light and fresh. Similar to taste, starting very sweet, with vanilla and apples; clean and smooth, with a scented finish, light bodied.

Glenburgie

REGION
Speyside

ADDRESS
Mains of Burgie, Forres,
Moray

PHONE
01343 850258

WEBSITE
No

OWNER
Chivas Brothers

VISITORS
By appointment

CAPACITY
4.2m L.P.A.

HISTORICAL NOTES: Glenburgie bore the name Kilnflat, a place in the parish of Alves, between Elgin and Forres, until the 1870s. It was built in 1829 by William Paul, the son of a distinguished surgeon, who had formerly been involved in Grange Distillery, close by and founded 1810 (hence the foundation date adopted by Glenburgie). In 1871, Paul sub-let to Charles Hay, who changed the name to Glenburgie and sold to Alexander Fraser & Company, in about 1882.

The new owner went bankrupt in 1925 and the receiver, a well-known character named Donald Mustard, took over but did not resume production and sold to the blending firm James & George Stodart Ltd of Dumbarton (owners of the Old Smuggler brand) two years later.

The giant Canadian distiller Hiram Walker bought 60% of Stodart's in 1930 – Walker's first move into Scotch – and took full control in 1936, when they also bought Miltonduff Distillery. They had acquired George Ballantine & Son Ltd the year before from James Barclay (see *Strathisla*), and Glenburgie and Miltonduff became (and remain) the key malts for the Ballantine's blends.

In 1987 Allied Lyons bought Ballantine's and its associated distilleries, and in 2005 they passed to Pernod Ricard and its operating division, Chivas Brothers.

Glenburgie Distillery was demolished in 2004 and rebuilt on an adjacent site at a cost of £4.3 million. The distillery came into production in June 2005, and next year an extra pair of stills were installed.

CURIOSITIES: Remarkably, the original distillery building of 1829 still stands, now isolated in a car park, but tastefully refurbished as a nosing room. It is tiny: a two-windowed, stone-built cottage, about 80 feet by 30, with an outside stair leading to a single room and a low-ceilinged cellar beneath.

Between 1958 and 1981, two Lomond stills were

The distillery stands on a slight eminence, the tail end of Burgie Hill, which was the 'blasted heath' where Macbeth met the witches.

in operation here, producing Glencraig single malt – named after William Craig, Ballantine's Production Director at the time. (See *Inverleven, Miltonduff.*) His son, who still works for the company, reported that the name was given to placate his father, who was sceptical about their effectiveness! The stills were converted to plain stills in the 1980s, by replacing the heads.

In his seminal book *Scotch Whisky* (1930), Aeneas Macdonald lists Glenburgie among his Top 12 Highland Malts.

The novelist Maurice Walsh was the Excise officer at Glenburgie during the 1940s. His books *The Quiet Man* and *Trouble in the Glen* were made into Hollywood movies; the first in 1950 (starring John Wayne and Maureen O'Hara), the second in 1954 (starring Orson Welles and Margaret Lockwood). His grandson is Jameson's master blender.

The views from the gallery in the malt loft across the Laich o' Moray to the Moray Firth are stupendous.

RAW MATERIALS: Unpeated malt from independent maltsters; own floor maltings removed 1958. Production water from springs on Burgie Hill; cooling water from a burn.

PLANT: Full-Lauter mash tun (eight tonnes); 12 stainless steel washbacks. Three plain wash stills (12,000 litres each charge); three plain spirit stills (14,000 litres each charge). Indirect fired by steam since 1958 (the riveted base of No. 1 wash still dates from the days of direct firing). Shell-and-tube condensers .

MATURATION: Mainly refill U.S. hogsheads, some first-fill. Dunnage, racked and palletised warehouses on site with capacity for 60,000 casks. The remainder matured at Mulben.

STYLE: Sweet, grassy and estery.

MATURE CHARACTER: The 15yo is light and grassy, with sweet vanilla sponge notes. The taste is sweet, with hints of vanilla, toffee and tinned pears, and a trace of smoke. Light-bodied.

Glencadam

REGION
Highland (East)

ADDRESS
Park Road, Brechin, Angus

PHONE
01356 622217

WEBSITE
www.glencadamdistillery.
co.uk

VISITORS
By appointment

OWNER
Angus Dundee plc

CAPACITY
1.3m L.P.A.

HISTORICAL NOTES: Situated about a mile outside the ancient Royal Burgh of Brechin, Glencadam (the name means 'the glen of the wild goose') was founded in 1825 by Messrs Thomas & Ruxton. In 1852 Alexander Milne Thompson took over, and in 1893, when Glencadam Distillery Ltd was incorporated, it was under the control of the Glasgow blenders Gilmore Thomson & Company. The directors of Gilmore Thomson sold the distillery to Hiram Walker for £83,400 in July 1954, the same year in which the Canadian company bought Scapa; Pulteney would be acquired in 1955.

When he visited in 1985, Philip Morrice noted: 'It is trimly furnished – in keeping with all the distilleries belonging to Hiram Walker – and is in no way dulled by its proximity to its nearest neighbour, the municipal cemetery.' Two years later Hiram Walker was bought by Allied Lyons (whose spirits division became Allied Distillers the same year). Glencadam became the heart-malt for Stewart's Cream of the Barley. Allied mothballed the distillery in 2000, then sold it to Angus Dundee Ltd, a well-established family firm of blenders based in London, in 2003. A large blending plant was installed at the distillery in November 2007.

CURIOSITIES: Glencadam has ascending lyne arms on both stills (by 14°), which increases reflux and makes for lighter spirit. Another uncommon feature is that the wash is heated externally in a heat exchanger attached to the wash still, then pumped back into the still where it is forced through a 'diffuser' to heat the rest of the wash. This greatly increases copper uptake so also makes for a lighter spirit.

Gilmore Thomson's 'Royal', for which blend Glencadam was a key malt, was a favourite of King Edward VII – a man who generally preferred champagne.

RAW MATERIALS: Unpeated malt from independent maltsters. Soft cooling and process water from Moorans water supply (a tributary of the North Esk), which used to supply the Royal Burgh of Brechin.

PLANT: Rake-and-plough mash tun (4.9 tonnes); six stainless steel washbacks. One plain wash still (12,600 litres); one plain spirit still (12,260 litres). Wash still has an external heater-diffuser. Indirect fired. Shell-and-tube condensers.

MATURATION: All ex-bourbon barrels. 20,000 casks on site.

STYLE: Soft, light, boiled sweets, pear drops.

MATURE CHARACTER: The 15YO has a pleasant 'peaches and cream' nose, with almonds and vanilla. The taste is sweet and creamy, with nuts and a trace of malt. Some detect asparagus and aniseed. Medium-bodied.

G1   Glendronach

REGION
Speyside

ADDRESS
Forgue, near Huntly,
Aberdeenshire

PHONE
01466 730202

WEBSITE
www.glendronach.com

VISITORS
Visitor centre

OWNER
The BenRiach Distillery
Company Ltd

CAPACITY
1.4m L.P.A.

HISTORICAL NOTES: Glendronach is one of the most charming and old-fashioned distilleries in Scotland. It was built in 1825 by a group of local farmers and businessmen led by James Allardice, who so impressed the local laird, the Duke of Gordon, that he was introduced by the Duke to London society, among whom he established something of a reputation for his 'Guid Glendronach'. Alas, a fire in 1837 largely destroyed the distillery, and this was followed five years later by Allardice's bankruptcy.

Most of his partners withdrew at this time, to be replaced by others who rebuilt the distillery during the 1850s (much of the present distillery dates from this time). The managing partner was Walter Scott, of Falkirk, who became sole proprietor in 1881. On his death in 1886, Glendronach was taken over by another partnership, this time of Leith wine merchants and a Campbeltown distiller. They managed it successfully until 1920 (even providing Teacher's Ardmore Distillery with its first Manager), when it was bought by Captain Charles Grant, youngest son of the founder of Glenfiddich Distillery (for £9,000).

His son sold to William Teacher & Sons in 1960, together with the distillery's 1,000-acre farm, and a herd of Highland cattle. The make had long been a key filling for the Teacher's blends. Glendronach was extended from two to four stills in 1966/67, then mothballed between 1996 and May 2002; and when Allied was split up in 2005, Teacher's went to Fortune Brands, and Glendronach Distillery to Pernod Ricard/Chivas Brothers.

On 25 July 2008, it was announced that the distillery had been sold to Billy Walker of BenRiach Distillers, who developed the brand massively. Then, on 27 April 2016, it was announced that the distillery, together with BenRiach and Glenglassaugh, had been bought for £285 million by the Brown-Forman

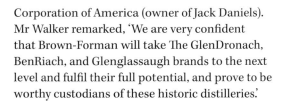

'Redcurrant jelly is good for the belly. Ginger and nuts are good for the guts. But the wine of Glendronach is good for the stomach!'
(traditional saying)

Corporation of America (owner of Jack Daniels). Mr Walker remarked, 'We are very confident that Brown-Forman will take The GlenDronach, BenRiach, and Glenglassaugh brands to the next level and fulfil their full potential, and prove to be worthy custodians of these historic distilleries.'

CURIOSITIES: On one occasion, when he had dined at Gordon Castle, Mr Allardice, somewhat the worse for drink, was over-effusive in his praise of the Duchess of Gordon's piano playing. The following morning the Duke informed him that his wife was not amused, to which he replied: 'Well, Your Grace, it was just the bottle of Glenlivet you gave me yesterday after dinner that did not agree wi' me. If it had been my ain guid Glendronach, I would have not have been ony the warr.' A cask of Glendronach was ordered immediately.

RAW MATERIALS: Production water from Dronac Burn, which flows through the distillery. Lightly peated malt from independent maltsters; own floor maltings, but not used since 1996.

PLANT: Unusually small cast iron, rake-and-plough mash tun (3.72 tonnes). Nine Oregon pine washbacks. Two boil-ball wash stills (9,000 litres charge); two boil-ball spirit stills (6,000 litres charge). Direct fired by coal until September 2005, when converted to indirect firing by steam. Shell-and-tube condensers.

MATURATION: Combination of ex-sherry and ex-bourbon casks.

STYLE: Rich, sweet, creamy.

MATURE CHARACTER: Richly sherried in style, but also showing the contribution (sweetness, vanilla, trace of coconut) of American oak. The taste is both sweet (with dried fruits, malt and toffee) and tannic-dry. Full- to medium-bodied.

G1 Glendullan

REGION
Speyside

ADDRESS
Dufftown, Moray

PHONE
01340 822100

WEBSITE
www.malts.com

VISITORS
By appointment

OWNER
Diageo plc

CAPACITY
5m L.P.A.

HISTORICAL NOTES: Glendullan Distillery was founded in 1897 by William Williams & Sons, blenders in Aberdeen, in a wooded glen beside the River Fiddich. Water from the river drove all the machinery in the distillery, via a huge water wheel – 'a great saving, compared to distilleries which have to use steam engines', wrote *Harper's*. It also shared a private railway siding with its neighbour, Mortlach. All supplies came by rail until 1968, when the branch line was closed.

The distillery was very well-built, and most of the original plant was still being used in the 1930s.

It was the seventh distillery built in Dufftown and gave rise to the rhyme: 'Rome was built on seven hills, but Dufftown stands on seven stills'.

In 1919 Williams amalgamated with Greenlees Brothers Ltd, Glasgow and London, makers of the Old Parr blends. This company had already merged with Alexander & Macdonald Ltd of Leith and now became Macdonald Greenlees & Williams. The company joined D.C.L. in 1925 and Glendullan was transferred to S.M.D. in 1930. The distillery was closed from 1940 to 1947, and rebuilt in 1962, the stills being converted to indirect heating at that time.

Ten years later a new and larger (six stills) distillery was built nearby, in S.M.D.'s 'Waterloo Street' style (see *Caol Ila* etc.). Between 1972 and 1985 the two distilleries operated in parallel, the makes of each being vatted together, then the original distillery was closed and dismantled. The buildings are now used as workshops by Diageo.

CURIOSITIES: Small amounts of Glendullan single malt were supplied to the Royal Household of Edward VII in 1902.

The spirit character from the original Glendullan Distillery was quite different from that of the 1970s distillery, and is esteemed more highly. The current spirit character is similar to Cardhu, a fact which

When Betty Boothroyd became Speaker of the House of Commons in 1992, she chose Glendullan as her 'Speaker's Choice'.

led to it being used in the infamous Cardhu Pure Malt (see *Cardhu*).

Since 2007 The Singleton of Glendullan has been popular in North America (see *Glen Ord*).

RAW MATERIALS: Floor maltings until 1962, now unpeated malt from Burghead. Soft process water from springs in the Conval Hills; cooling water from the River Fiddich. Previously, Fiddich water was used for mashing as well – one of the reasons for choosing the site was its proximity to the river.

PLANT: Full-Lauter mash tun (12 tonnes). Three plain wash stills (15,500 litres charge); three plain spirit stills (16,000 litres charge). Until 1972, two stills only. All indirect fired. Shell-and-tube condensers. The original distillery had worm tubs, in use until 1985.

MATURATION: Ex-bourbon refill hogsheads, some European oak (for The Singleton). Matured in the Central Belt

STYLE: Floral/grassy.

MATURE CHARACTER: Typical light/medium-style Speyside. The nose is sweet and estery, with apples and pears and cut grass; the taste is sweet throughout with a smooth texture and light mouthfeel. Medium- to light-bodied.

Glen Elgin

REGION
Speyside
ADDRESS
Longmorn, Elgin, Moray
PHONE
01343 862100
WEBSITE
www.malts.com
OWNER
Diageo plc
VISITORS
By appointment
CAPACITY
2.7m L.P.A.

HISTORICAL NOTES: Charles Doig, the famous distillery architect who designed Glen Elgin, prophesied that it would be the last distillery to be built on Speyside for 50 years. He was spot on: the next was Glen Keith, which opened in 1958.

Glen Elgin was founded by a former Manager at Glenfarclas, William Simpson, in partnership with a banker, James Carle. Building started in 1898, but following the crash of Pattison's – a major blending house and buyer of fillings – it was down-sized, and within six months of its opening in 1900 it closed and was sold for £4,000. Production resumed for a year, then Glen Elgin was sold again in 1906 (for £7,000) to John J. Blanche & Company, a wine and spirit company in Glasgow. Blanche died in 1929 and again the distillery was sold, this time to D.C.L., who licensed it to White Horse Distillers – it had long been a key constituent of the White Horse blend.

Until the 1950s the distillery was entirely operated and lit by paraffin; all machinery was driven by a paraffin engine and water turbine. Like other D.C.L. distilleries, Glen Elgin was extensively refurbished in 1964, and expanded from two stills to six; it closed again for refurbishment between 1992 and 1995.

Glen Elgin is ranked Top Class by blenders, and it has been described as 'the distillers' dram of drams'.

CURIOSITIES: Brian Spiller remarks that: 'It took almost the whole of one man's time to keep the paraffin lamps in working order.' Glen Elgin was only connected to the electricity grid in 1950.

RAW MATERIALS: Floor maltings until 1964, now unpeated malt from Burghead. Soft process water from springs near Millbuies Loch; cooling water from the Glen Burn.

PLANT: Full-Lauter tun, with an octagonal canopy (8.2 tonnes). (The earlier rake-and-plough mash tun was only removed in 2000.) Six larch washbacks. Three plain wash stills (6,800 litres charge); three plain spirit stills (8,100

Glen Elgin's site was chosen for the quality of its water supply and proximity to the railway. Unfortunately, the water source proved unreliable and permission for a railway siding was refused by the Board of Trade!

litres charge). All indirect fired (direct fired by coal until 1970). Worm tubs.

MATURATION: Mix of ex-bourbon and ex-sherry refill casks.

STYLE: Fruity and full-bodied.

MATURE CHARACTER: Glen Elgin is subtle and complex. At first sight it is a typical Speyside – estery, fruity, grassy – but there is interesting depth to these aromas, traces of tangerine, light honey, vanilla, even a hint of cloves. Medium-bodied.

Glenesk Dismantled

REGION
Highland (East)
ADDRESS
Kinnaber Road, Hillside,
Montrose, Angus
LAST OWNER
D.C.L./S.M.D.
CLOSED
1985

No distillery has been known by so many names!

HISTORICAL NOTES: When it was founded in 1897, within a converted flax mill on the bank of the River South Esk, it was named 'Highland Esk'. By 1897 it was 'North Esk'; between 1938 and 1964 it was called simply 'Montrose'; then its name changed to 'Hillside' (1964); finally it became 'Glenesk' in 1980.

The mill conversion was done by James Isles, a wine merchant in Dundee, in partnership with Septimus Parsonage & Company. They lasted only two years, when the distillery was taken over. Closed during World War I, it did not operate again until 1938, when it was bought by Joseph Hobbs for National Distillers of America (see *Ben Nevis*). A patent still was installed and Glenesk/Montrose was converted to grain distillation. When D.C.L. bought it in 1954 they continued to distill grain whisky intermittently, then resumed malt whisky distilling in 1964.

A drum maltings was built beside the distillery in 1968 (enlarged in 1973 to 24 drums) – Montrose's situation is on the edge of the fertile Mearns district, good barley-growing country. The maltings was bought by Pauls Malt Ltd in 1996, and is now owned by the Irish company Greencore, the sixth largest malt producer in the world.

The distillery closed in 1985, although the distilling licence was only cancelled in 1992. All the distilling equipment has now been removed.

CURIOSITIES: 'Esk' derives from *uisge*, the Gaelic for 'water'.

Glenesk was licensed to William Sanderson & Sons during its D.C.L. days, and supplied fillings for VAT 69.

Glenfarclas

REGION
Speyside
ADDRESS
Ballindalloch, Moray
PHONE
01807 500257
WEBSITE
www.glenfarclas.co.uk
OWNER
J. & G. Grant
VISITORS
Visitor centre, shop
(with panelling from
RMS *Empress of Australia*)
CAPACITY
3.5m LPA

HISTORICAL NOTES: The distillery was not actually founded by the Grants of Glenfarclas: the first licence for the site was granted in 1836 to Robert Hay of Rechlerich Farm (there had been an unlicensed enterprise here since 1797), and when he died in 1865 his neighbour John Grant bought the distillery for £512. He leased it to John Smith, who had managed The Glenlivet Distillery, went on to build Cragganmore, and was reckoned one of the best distillers of the day.

In 1896, a second pair of Grants, John and George (grandsons of the founder), took over and rebuilt the distillery.

It was expanded in 1960 by the third pair of J. & G. Grants (great-grandsons, who took over in 1949), when a second pair of stills was added; this was increased to three pairs in 1976. George S. Grant served 52 years as Chairman of the family company (1949–2001), when he was succeeded by his son John. John's son is now Global Ambassador for Glenfarclas, and set to take over one day . . . His name? Why, George, of course!

CURIOSITIES: 'Glenfarclas' means 'the valley of the green grass'; it stands in meadows at the foot of Ben Rinnes. The distillery has the largest stills on Speyside and was the first to release a cask strength single malt, in 1968, later named '105' (i.e. five degrees over proof).

It was also one of the first distilleries to build a visitor centre, in 1973, which included a tasting room fashioned from the stateroom of RMS *Empress of Australia* (built 1913–1919, broken up 1952).

Among the company's core and (many) other expressions is the exceptional range of old malts (1952–1998), The Family Casks. This is being added to annually, and in some years cask variations are available: for example, 1990s bottlings are from first-fill Oloroso, first-fill Fino and first-fill ex-bourbon casks.

'Of all the whiskies, malt is king – of all the kings, Glenfarclas reigns supreme!' (1912, quoted by a rival distiller!)

RAW MATERIALS: Floor maltings until 1972, now lightly peated malt from independent maltsters. Soft process and cooling water from springs above the snow line on Ben Rinnes.

PLANT: Largest semi-Lauter mash tun on Speyside (16.5 tonnes per mash). Twelve stainless steel washbacks. Three large boil-ball wash stills (25,000 litres charge), equipped with rummagers; three boil-ball spirit stills (21,000 litres charge). All direct fired by gas (with oil as back-up). Shell-and-tube condensers.

MATURATION: Now mainly Oloroso ex-sherry casks, used up to three times. 60,000 casks matured on site, with about a third refill bourbon (so-called 'plain' casks).

STYLE: Delicate, sweet and fruity – gains weight during maturation.

MATURE CHARACTER: Glenfarclas ages very well, especially in sherry-wood. Above 15yo it develops complexity which is not apparent in the younger expressions – combining sherry, fruit cake and orange marmalade with sweet malt, nuts and tannic dryness, in a wholly satisfactory way, and rarely with the sulphury notes often associated with sherry-wood.

Glenfiddich

REGION
Speyside

ADDRESS
Dufftown, Moray

PHONE
01340 820373

WEBSITE
www.glenfiddich.com

OWNER
William Grant & Sons Ltd

VISITORS
Reception centre,
large shop, conference
facilities, restaurant

CAPACITY
13m L.P.A.

*For many years,
Sir Compton
Mackenzie (author
of the bestseller
Whisky Galore!)
featured in the
print advertising for
Grant's Standfast.*

HISTORICAL NOTES: In the autumn of 1886 William Grant, son of a Dufftown tailor and Manager of Mortlach Distillery, bought the distilling equipment from Elizabeth Cumming of Cardhu for £120 (including stills and a water mill). Since his annual salary was £100, it had taken him many years to amass enough to go it alone. With his wife and nine children he set about carting stone from the bed of the River Fiddich and building his distillery on a site on the edge of the town named Glenfiddich, 'the valley of the deer'. The first whisky ran from its stills on Christmas Day 1887.

His family all joined William Grant in the enterprise and the company is still controlled by his descendents, now the fifth generation. Fortunately, soon after Glenfiddich began production, William Williams & Sons, blenders in Aberdeen, placed an order for 400 gallons a week – the entire output of the nascent distillery. Soon the family were offering their own blends, including Standfast (the motto of Clan Grant), and selling them overseas as well as in the U.K. By 1914 the company had established 63 agencies around the world.

In 1963 the Directors of William Grant & Sons took the unprecedented step of bottling Glenfiddich Pure Malt (Straight Malt in the U.S.A.) and marketing it in the same way as blended Scotch had always been marketed; at first in England, then overseas. The venture was a huge success, export sales alone rising from 4,000 cases in 1964 to 119,500 cases in 1974. That year the company was granted the Queen's Award for Export Achievement, the first whisky company to be so honoured.

CURIOSITIES: In spite of having the largest number of pot stills in the world, the distillery is traditional in many ways. All the stills are direct fired (by gas since 2003, previously by coal); it has its own cooperage and coppersmiths; maturation is done on site;

For many years Glenlivet was the best-selling single malt in the world, commanding around 18% of the market.

and until recently all bottling was also done at the distillery (from 2007 the bottling of 12YO has been done at Bellshill, near Glasgow, although the spirit is still reduced with process water from the Robbie Dubh spring).

Glenfiddich was the first distillery to open a visitor centre, in 1969. Today it has five stars from VisitScotland and a summer staff of around 40, speaking 11 languages and welcoming visitors from between 105 and 110 countries. Grant's have invested £2 million in the visitor facilities in recent years and Glenfiddich offers a range of tours at different prices.

Unveiled in spring 2008, Glenfiddich commissioned a magnificent life-size bronze stag from the eminent sculptor Tessa Campbell Fraser. The red deer stag has long been the emblem of the distillery: 'fiddich' derives from the Anglicisation of the Gaelic *fiadh*, meaning deer.

The unique triangular-shaped bottle commissioned by Grant's in 1956 for their Standfast blend, and later used for Glenfiddich single malt, was designed by Hans Schleger. Ever innovative, the family saw the opportunity offered by the emergence of duty-free shops in airports (the first opened in Shannon around 1968). As a result, English tourists' first encounter with malt whisky was often while on holiday, which gave it happy associations.

Another simple but effective promotion was to offer Standfast bottles filled with tinted water as stage props. As a result it became a common sight in TV dramas of the 1960s!

Sales of Glenfiddich topped one million cases in 2011, and turnover passed £1 billion in 2012, but it lost its position as the best-selling malt whisky in the world to The Glenlivet in 2014.

RAW MATERIALS: Floor maltings until 1958, now unpeated malt from independents. Soft process water from the Robbie Dubh springs.

PLANT: Two full-Lauter mash tuns (10.2 tonnes per mash, eight mashes daily. Glenfiddich was the first distillery to employ continuous mashing). Twenty-four Oregon pine washbacks. Ten plain wash stills (charge 9,500 litres); 18 boil-ball spirit stills (charge 5,500 litres). All direct fired by gas. Shell-and-tube condensers.

MATURATION: On site, in a mix of American and European oak (15% European/85% American in Special Reserve).

STYLE: Light Speyside. Floral and fruity.

MATURE CHARACTER: Classic Speyside style – light, fresh, fragrant, fruity. The nose adds cereal notes and a pine-sap dimension, and sometimes I detect a very light scent of coal smoke. The taste is sweet throughout, with fresh citric notes. Supremely accessible. Light-bodied.

Glengarioch

REGION
Highland (East)

ADDRESS
Oldmeldrum,
Aberdeenshire

PHONE
01651 873450

WEBSITE
www.glengarioch.com

OWNER
Beam Suntory

VISITORS
Visitor centre opened
2006, with shop, V.I.P.
tours and small confer-
ence centre

CAPACITY
1.37m L.P.A.

HISTORICAL NOTES: The Garioch (pronounced 'Geery') is the fertile tract of arable land, 150 square miles in size, at the heart of Aberdeenshire, formerly known as that county's 'granary'. The market town of Oldmeldrum stands within it, and the current distillery was built here in 1797, possibly on the site of the earlier Meldrum Distillery which dates from before 1785. The founder was John Manson, joined four years later by his son, Alexander.

It changed hands 40 years later, and again in 1884, when it was bought by J.G. Thomson & Company of Leith, an old established firm of wine and spirits merchants whose offices, The Vaults, now house the Scotch Malt Whisky Society. William Sanderson, creator of VAT 69, became 'proprietor' around 1908.

In 1933 William Sanderson & Son merged with Booth's Distilleries Ltd, owners of Royal Brackla, Millburn and Stromness Distilleries as well as their eponymous gin, and four years later the amalgamated company joined D.C.L. Management of Glengarioch was undertaken by S.M.D. in 1943; in 1968 they mothballed the old distillery – on account of 'chronic water shortages and limited production potential' – and two years later sold it to the Glasgow whisky broker (and owner of Bowmore Distillery) Stanley P. Morrison.

Having solved the water problem by digging a deep well in a nearby field, Morrison extended the plant from two to three stills in 1971, and to four in 1973, but retained the floor maltings. Peating levels were increased, the peat coming from New Pitsligo Moss nearby. The maltings were closed in the early 1990s, at which time the stills were converted to indirect firing by steam coils and pans. Morrison Bowmore Distillers were taken over by Suntory in 1994, and since then Glengarioch has endured periods of closure. A visitor centre was opened in January 2006, and was refurbished in early 2011. The whiskies were elegantly repackaged in 2009.

The distillery is named Glengarioch, but its product Glen Garioch.

CURIOSITIES: During the 1970s, fuel costs escalated (rising from 9% of production costs to 16% by 1980). Morrison's installed an innovative waste heat recovery system at Glengarioch which not only supplied heat for the kiln and pre-heated the wash, but provided heating to two acres of glasshouses, saving around £90,000 per annum. The distillery became famous for its tomatoes and hothouse plants, but the practice was discontinued in 1993.

On another occasion that water was being sought, after the owning company, Morrison Bowmore, had been taken over by Suntory, the distillery employed a famous water diviner, Neil Murchie, known as 'The Witch Mannie o' Foggieloan'. Neil spoke in broad Doric, difficult even for Scots to understand, and impossible for the translator who was trying to interpret for a Japanese representative of the company!

RAW MATERIALS: Floor maltings until 1993, supplying half the distillery's requirement; now unpeated malt from Simpson's of Berwick-upon-Tweed. Soft process water from Coutens Spring; cooling water from there and Meldrum Burn.

PLANT: Stainless steel full-Lauter tun with a peaked canopy (four tonnes per mash). Six stainless steel washbacks, one plain wash still (charge 21,500 litres), one plain spirit still (charge 9,500 litres). Both direct fired by gas until 1995, now indirect fired by steam. Shell-and-tube condensers, with after-coolers until 1993.

MATURATION: Bourbon and sherry casks. Four dunnage warehouses on site, with a capacity of 12,000 casks.

STYLE: Medium-bodied, slightly fruity/estery.

MATURE CHARACTER: Medium-rich in style, beefed up by the use of sherry-wood. A curious lavender note on the nose, alongside sherry and malt, and a distinct whiff of smoke. Sometimes also ginger snaps. The taste combines toffee-sweetness with some tannic dryness and again a hint of smoke in the finish.

Glenglassaugh

REGION
Speyside

ADDRESS
Portsoy, Aberdeenshire

PHONE
0131 335 5130

WEBSITE
www.glenglassaugh.com

OWNER
The BenRiach Distillery
Company Ltd

MOTHBALLED
1986

VISITORS
Visitor centre planned

CAPACITY
1.1m L.P.A.

HISTORICAL NOTES: Glenglassaugh was the distillery that its owners didn't talk about! It was built on the outskirts of the ancient burgh and harbour of Portsoy, on the Banffshire coast, in 1874 by an enterprising local businessman, James Moir, in partnership with two of his nephews and Thomas Wilson, coppersmith. Their intention was to sell most of their make as 'self' or single malt whisky, and they found a ready market for the surplus in Robertson & Baxter, brokers and blenders in Glasgow.

Moir died in 1887; his nephew refurbished the distillery, but when his brother died in 1892 he decided to sell Glenglassaugh to pay death duties. He offered it to Robertson & Baxter, who immediately sold it on to their sister company, Highland Distilleries (now both part of the Edrington Group). Demand for Glenglassaugh dwindled after 1898; the distillery was silent from 1907 to 1960, when it was refurbished. It closed again in 1986, and, apart from a short period in production in 1998, it remained silent. The warehouses on site were used by Edrington.

In February 2008, it was announced that Edrington had sold Glenglassaugh to a Dutch-based consortium, the Scaent Group, for £5 million. Stuart Nickerson, a former Distilleries Director with William Grant & Sons, who advised the consortium in their search for a distillery, was appointed Managing Director.

The restored distillery was opened on 24 November 2008 by the local M.S.P., Alex Salmond, then the First Minister of Scotland. An elegant and comprehensive book about Glenglassaugh by Ian Buxton was published in 2010.

In 2013, ownership changed again when it was bought by Billy Walker of BenRiach and Glendronach Distilleries, and in 2016 all three distilleries were sold for a massive £285 million to Brown-Forman, of Louisville, Kentucky.

The charming harbour of Portsoy was considered to be the safest on the north-east coast in the sixteenth century, although a new harbour built in 1825 was swept away by a storm.

CURIOSITIES: James Moir sold seeds, ironmongery, manure, wines and spirits. He was agent for the North of Scotland Bank, owner of a fishing boat and of salmon nets on the Deveron. He was also (after 1865) Colonel of the Local Volunteer Artillery Regiment.

In the mid 1980s Highland Distilleries wanted to increase the sweet, Speyside style of Glenglassaugh for use in The Famous Grouse. It was believed at the time that soft water might help achieve this. Glenglassaugh's water was hard. After tankering some soft water in from Glen Rothes for tests, an experimental water softening plant was installed and then a full plant. This proved too costly, and led to the decision to mothball Glenglassaugh and expand Glen Rothes.

RAW MATERIALS: Medium-hard water from two deep wells in the vicinity of the Glassaugh Burn, which previously was softened on site prior to mashing. This softening process is not used now. Unpeated malt from commercial maltsters.

PLANT: Copper covered cast iron mash tun with traditional stirring gear and old-fashioned underback (5.25 tonnes mash), two stainless steel and four wooden washbacks. One boil-ball wash still (11,000 litres charge), one boil-ball spirit still (12,000 litres charge). Both indirect fired by steam pans. Shell-and-tube condensers.

MATURATION: Previously in a variety of ex-bourbon barrels, ex-sherry butts and refill hogsheads, matured on a variety of sites.

STYLE: Fruity, sweet, with traces of smoke, and spice, and an unusually dry and salty finish.

MATURE CHARACTER: Sweet, with orange juice and pears, light oil and sea salt and sometimes traces of 'dried shellfish'.

Glengoyne

REGION
Highland (South)

ADDRESS
Dumgoyne, by Killearn,
Stirlingshire

PHONE
01360 550254

WEBSITE
www.glengoyne.com

OWNER
Ian Macleod Distillers

VISITORS
Visitor centre, blending
courses and its own
helipad

CAPACITY
1.1m L.P.A.

HISTORICAL NOTES: Glengoyne straddles the
Highland Line, with its warehouses below, and
the distillery itself above. Until the 1970s it was
classified as a Lowland malt. The land hereabouts
is owned by the Edmonstones of Duntreath, and it
was a representative of this family who obtained a
licence to distill in 1833, under the name 'Burnfoot'.

The pretty site is indeed at the foot of the fast
flowing Blairgar Burn, in a steep-sided wooded glen,
covered in bluebells in spring.

The distillery was leased to George Connell,
John McLelland (1851–67) and Archibald McLellan
(1872–76), then to Lang Brothers, whisky blenders in
Glasgow, at which time it was renamed Glen Guin.
Alexander and Gavin Lang commenced trading in
1861 from the basement of the Argyll Free Church
in Oswald Street (which they later took over as a
bonded warehouse), giving rise to the jingle: 'The
spirits below were the spirits of wine and the spirits
above were the spirits Divine.'

Lang Brothers had long bought fillings from
Robertson & Baxter, and in 1965 became wholly
owned by R. & B. The distillery was refurbished and
the stills increased from two to three. R. & B. was
consolidated into the Edrington Group in 1999, and
in 2003 Lang Brothers Ltd and Glengoyne Distillery
were sold to Ian Macleod & Company, whisky
blenders, of Broxburn.

Glengoyne offers the widest range of visitor
experiences, including a masterclass during which
you create your own vatting of single malt from
different cask styles. In 2015 the distillery hosted
80,000 visitors.

CURIOSITIES: Air Marshall Sir Arthur Tedder, first
Baron Tedder of Glenguin, was born at the distillery,
where his father was the Excise officer from 1889
to 1893. Arthur (Senior) became Chief Inspector
of Excise, and was knighted for his services to the
Royal Commission of Enquiry into Whisky (1909).

'Glengoyne' derives from 'Glen Guin', 'the glen of the wild geese'. The present name was adopted in 1905.

RAW MATERIALS: Soft water from Blairgar Burn which runs from the Campsie Hills. Floor maltings until 1910; now malt from Simpson's, Berwick-upon-Tweed.

PLANT: Traditional rake-and-plough mash tun with copper canopy (3.72 tonnes), six Oregon pine washbacks, one boil-ball wash still (charge 14,000 litres), two boil-ball spirit stills (charge 3,495 litres per still); all indirect fired. Shell-and-tube condensers.

MATURATION: Sherry (around 40%) and refill casks. Matured on site in three dunnage warehouses (6,700 casks), with plans to build a further two racked warehouses.

STYLE: Light fruity, hint of vegetal due to and mash.

MATURE CHARACTER: Glengoyne gains weight with age. The core bottlings are a mix of ex-sherry and ex-bourbon casks. The nose is malty with sherry notes and a bruised-pear fruitiness. The taste is well-balanced: light sweetness, some acidity, drying in the finish. Medium-bodied

G1   Glen Grant

REGION
Speyside

ADDRESS
Elgin Road, Rothes,
Moray

PHONE
01340 832118

WEBSITE
www.glengrant.com

OWNER
Glen Grant Ltd
(Campari Group)

VISITORS
Visitor centre, with shop
and extensive woodland
garden

CAPACITY
6.2m L.P.A.

HISTORICAL NOTES: When their lease at Aberlour expired in 1839 the brothers John and James Grant moved down the road to Rothes and the following year built a new distillery (originally named Drumbain), the first distillery in this village. From the outset it was described as 'one of the most extensive distilleries in the North', although goods had to be transported by road until the arrival of the railway in 1858 – a project in which James Grant (by then Provost of Elgin) was closely involved. An early engine was named *Glen Grant*.

James Grant Junior, known as 'the Major', succeeded his father in 1872, at the beginning of the Whisky Boom. Glen Grant was already being sold as a single, described as 'pure, mild and agreeable . . . peculiarly adapted for family use', in 'England, Scotland, and the Colonies'. It was sold by the cask, but the distillery provided customers with its distinctive label, featuring two kilted Highlanders seated beside a butt, with the motto 'From the Heath Covered Mountains of Scotia I Come'.

By 1887, the demand for Glen Grant was such that the Major built another distillery nearby, Glen Grant Number Two (see *Caperdonich*), and in 1898 installed the first pneumatic malting drums in the Highlands – electrically driven, and in operation until 1971, when on-site malting was discontinued.

The Major lived the life of a Victorian laird in Glen Grant House beside the distillery (now demolished), and laid out the splendid woodland garden that extends behind the distillery. In his day, it required 15 gardeners to keep it neat. It was restored and opened to the public in 1996, complete with the small safe in a summer house from which the Major would produce a bottle of Glen Grant for his guests! He died in 1931, aged 84, and was succeeded by his grandson, Douglas Mackessack.

In 1952, Glen Grant merged with George & J.G. Smith Ltd, to form The Glenlivet & Glen Grant Distilleries Ltd; in 1970 this concern amalgamated

Glen Grant was one of the very few single malts widely available pre-1970.

with the blending house Hill, Thomson & Company, and with Longmorn Distilleries, to become The Glenlivet Distillers Ltd, which was purchased by The Seagram Company in 1978. Douglas Mackessack announced this 'with regret, but as being inevitable if Glen Grant's economic survival was to be assured', handed a cash bonus to all employees and retired.

In 2001, Pernod Ricard bought Seagram's Scotch whisky interests and were obliged by E.U. monopolies regulations to sell Glen Grant. It was bought by Campari of Milan in 2006, who appointed Dennis Malcolm as Distillery Manager. He joined Glen Grant in 1961, becoming Manager in 1981 and later General Manager for all the Seagram distilleries in 1992.

CURIOSITIES: The heating of the stills at Glen Grant is curious. The original four stills were direct fired by coal, the wash stills having rummagers driven by a water wheel (the last in the industry, used until 1979). Two more were added in 1973, direct fired by gas (L.P.G.), and a further four in 1977. In 1979 all ten were converted to gas. But this regime only lasted until 1983, when again the wash stills reverted to coal, using a waste heat boiler to heat steam for the (indirect fired) spirit stills. At this time the old stillhouse (containing the four original stills) was replaced by a new stillhouse, with two larger stills, so capacity was unaffected. All were converted to indirect firing in the late 1990s.

The so-called 'German helmet' wash still design, which has the appearance of a giant's hand-bell, is unique to Glen Grant. The four wash stills all have this peculiarity, and all are equipped with purifying drums on their necks to increase reflux.

The Glen Grant range was repackaged in 2007, the label stressing (twice!) 'Distilled in Tall Slender Stills'.

Thanks to the entrepreneurial flair of the distillery's Italian agent, Armando Giovinetti,

'The character
of whisky is
determined not
by the purity
of the spirit
manufactured, but
by the impurities
left in the spirit.'
Douglas
Mackessack

Glen Grant became the first single malt to take off in an export market. Giovinetti realised that a young expression would have most appeal (he was challenging Grappa), and specified five years old. He promoted this vigorously from the mid 1960s, and by 1977 was selling around 200,000 cases a year.

Glen Grant is still Italy's favourite malt, although sales have declined in recent years. This has been compensated for by increased activity in France and the U.S.A. The brand sells over three million bottles worldwide, and is among the world's best-selling malt whiskies.

RAW MATERIALS: Unpeated malt from independent maltsters (own maltings removed 1962). Water from the Caperdonich Springs and the Glen Grant Burn (formerly named the Black Burn).

PLANT: Semi-Lauter mash tun (12.28 tonnes); ten Oregon pine washbacks. Four wash stills (15,100 litres charge each), of 'German helmet' design; four boil-pot spirit stills (7,800 litres charge each). Direct fired until late 1990s, now all indirect fired by steam; all fitted with purifiers. Worm tubs until 1980s, now all shell-and-tube condensers, with after-coolers and hot water recovery devices for pre-heating the charges to all eight stills.

MATURATION: Mainly refill U.S. hogsheads, mostly on Chivas sites (especially at Mulben, near Keith).

STYLE: Sweet, grassy and fruity (green apples).

MATURE CHARACTER: The house style favours refill wood to produce a light-coloured, fresh, fruity and summery whisky. The taste is sweet and lightly lemony, with cereals and nuts, apples and pears. Light-bodied.

Glengyle

REGION
Campbeltown

ADDRESS
Glengyle Road,
Campbeltown, Argyll

PHONE
01586 551710

WEBSITE
www.kilkerransingle-
malt.com

OWNER
Mitchell's Glengyle Ltd

VISITORS
No

CAPACITY
750,000 l PA

HISTORICAL NOTES: The original Glengyle Distillery operated from 1873 to 1925. It was built by William Mitchell & Company and remained in their ownership until 1919, when it was sold to West Highland Malt Distilleries Ltd. The distillery was closed in 1925; the stock and warehouses were sold – the latter became the Campbeltown Miniature Rifle Club, and was bought by Bloch Brothers in 1941 (see *Glen Scotia*). They announced that they were rebuilding and expanding, and even installing a grain distillery at Glengyle, but nothing happened and the site was sold to Argyll Farmers and used as a depot and sales office by Kintyre Farmers Co-operative.

It was bought by J. & A. Mitchell Ltd, owner of Springbank Distillery, in November 2000; a new distillery was built and production commenced in March 2004. Currently production is limited by arrangements with Springbank.

CURIOSITIES: William Mitchell, the founder of the original Glengyle Distillery, was the son of Archibald Mitchell, one of the original partners in J. & A. Mitchell, who rebuilt the distillery. Hedley Wright, the Chairman of J. & A. Mitchell Ltd, is his great-great nephew.

Springbank and Glengyle produce all their malt requirements on site, in floor maltings.

RAW MATERIALS: Water from Crosshills Loch. Lightly peated malt from Springbank Distillery maltings.

PLANT: Stainless steel semi-Lauter mash tun with a canopy (four tonnes). Four larch washbacks. One plain wash still (second-hand from Ben Wyvis Distillery; 11,000 litres charge), one plain spirit still (9,000 litres charge). Both indirect fired. Shell-and-tube condensers.

MATURATION: Bonded warehouses at Springbank Distillery.

STYLE: Heavier than Springbank, oily, meaty, sweet with cereal notes.

G1   Glen Keith

REGION
Speyside

ADDRESS
Station Road, Keith,
Banffshire

PHONE
01542 783042

WEBSITE
No

OWNER
Chivas Brothers

CAPACITY
6m L.P.A.

HISTORICAL NOTES: Glen Keith is situated on the opposite bank of the River Isla to Strathisla Distillery in the town of Keith. It was the first of Seagram's foundations, having been created out of a former meal mill in 1957/58, and was licensed to their subsidiary, Chivas Brothers. It was also the first distillery to have been built on Speyside since 1900, and, as Philip Morrice writes, 'an excellent job has been done in recreating the form and ambience of a turn-of-the-century Highland malt whisky distillery'. Although a modern unit, Glen Keith is traditional in design and partly built of dressed stone, including its pagoda-topped kiln.

It was originally designed for triple distillation, until 1970, when the number of stills increased from three to five (the sixth still was installed in 1983). These were the first stills in Scotland to be direct fired by gas, although they became indirect fired by steam pans and coils only three years later. The gradual expansion of capacity is reflected in the different shapes and sizes of stills, and a number of production trials have been run at Glen Keith over the years.

Seagram's mothballed Glen Keith in 1999, and the company's Scotch whisky interests were acquired by Pernod Ricard in 2001.

In 2012/13 the old Saladin maltings were demolished and a new building raised on the site, equipped with a highly efficient Briggs full-Lauter mash tun and six stainless steel washbacks. New malt storage has also been added; the existing nine wooden washbacks in the original building have been replaced, and the six stills upgraded. Capacity has increased from 4m to 6m L.P.A.

CURIOSITIES: Seagram's laboratory, now the Chivas Technical Centre, is based at Glen Keith and services the needs of the entire group. The distillery was also formerly the 'brand home' of Passport blended Scotch whisky.

RAW MATERIALS: Own Saladin maltings until 1976; now unpeated malt from independent maltsters. Soft process and cooling water from Balloch Spring, augmented when required from Newmill Spring.

PLANT: Full-Lauter mash tun (8 tonnes). Six stainless steel washbacks and nine Oregon pine washbacks. Three plain wash stills (15,300 litres charge each), two boil-ball spirit stills, one plain spirit still (11,100 litres charge each). Indirect fired by steam coils and pans.

MATURATION: Mainly refill casks, at various Chivas sites on Speyside and in the Central Belt.

STYLE: Light, sweet, grassy

MATURE CHARACTER: Speyside style, with apples and bananas, hedgerow flowers, lemongrass and vanilla. Sweet to taste, with dried fruits (figs, dates) and light almonds. Medium-bodied.

Glenkinchie

REGION
Lowland

ADDRESS
Pencaitland, East Lothian

PHONE
01875 342012

WEBSITE
www.malts.com

OWNER
Diageo plc

VISITORS
Visitor centre, with
museum and shop

CAPACITY
2.5m L.P.A.

HISTORICAL NOTES: The original distillery was called Milton, a typical farm distillery, established by the brothers John and George Rate in 1825 to make use of the surplus grains they produced after adopting the improved farming methods introduced by the Agricultural Revolution. Indeed, their landlord may have been John Cockburn of Ormiston, 'The Father of Scottish Husbandry' and one of the pioneers of agricultural improvements.

In 1837 they moved to the present site and built a new distillery, but by 1852 they were bankrupt. Their successor used part of the premises as a sawmill, but in 1880 the site was bought by a consortium of brewers and whisky merchants in Edinburgh, and production resumed. By 1890 the owner was Major James Grey, who rebuilt the distillery and associated buildings as a model village. In 1914 he joined four other Lowland distillers to form Scottish Malt Distillers, a trade relationship designed to consolidate the interests of malt distillers in the Lowlands during a time of recession. This body joined D.C.L. in 1925, and Glenkinchie was licensed to John Haig & Company.

In 1988 Glenkinchie was adopted by the recently founded United Distillers plc as the Lowland representative in their Classic Malts series.

CURIOSITIES: The name comes from the Kinchie Burn, which supplies cooling water, and this name is a corruption of de Quincey, the family which owned the lands in medieval times.

In 1895 a fire started in the stables during an extremely cold night – so cold that the water froze in the hoses, and the staff were powerless. This so upset the distillery brewer that he went mad and was consigned to an asylum.

Glenkinchie had a substantial farm attached, and was long famous for its beef cattle, fattened on the residues of the process. The post-war Distillery Manager, Mr W.J. McPherson, won Supreme

Champion at Smithfield show in 1949, 1952 and 1954. For many years James Buchanan & Company's famous Clydesdale dray-horses spent their summer holidays at Glenkinchie.

With a capacity of nearly 21,000 litres, Glenkinchie's wash still is the largest pot still in Scotland. A large bronze bell bearing the date 1842 hangs in the mashhouse. Until 'dramming' ceased at the distillery, the bell was sounded to mark the end of each shift – the time for drams to be dispensed!

The distillery's excellent visitor centre also houses what used to be called 'The Museum of Malt Whisky'. This was begun by a former Manager, then taken up by S.M.D. It was redesigned and reopened in 1995. At its centre is a scale model (1:6) of a malt whisky distillery designed in 1924 for the British Empire Exhibition at Wembley by James Cruikshank, S.M.D.'s General Works Manager, and showing the whole process from malting to maturation. Adjacent to the distillery is a neat bowling green, installed about 1900 for the entertainment of distillery staff.

RAW MATERIALS: Floor maltings until 1968; now lightly peated malt from Roseisle. Hard process water from a spring on site (formerly from Hopes Reservoir in the Lammermuir Hills); cooling water from the Kinchie Burn.

PLANT: Full-Lauter tun (9.4 tonnes). Five Oregon pine washbacks and one of Douglas fir. One lamp-glass wash still (20,000 litres charge); one lamp-glass spirit still (17,000 litres charge). Indirect fired since 1972. Worm tubs – uniquely, the worm on the spirit still is made from stainless steel (which increases the sulphury character required at Glenkinchie). The lye pipe connecting the worm to the still dips steeply, like an elephant's trunk, as it leaves the stillhouse. This also inhibits reflux and enhances heaviness.

MATURATION: Mainly ex-bourbon casks, matured on site and at Leven.

STYLE: Heavy and meaty, with sulphury notes.

MATURE CHARACTER: During maturation, the heavy style of Glenkinchie's new-make becomes fresh and fragrant, and it loses all traces of sulphur. The nose is 'rural' – meadows, hedgerows, with lemon notes and a thread of smoke. The taste is fresh, starts lightly sweet and finishes dry, and very short. Light in character, but with medium body.

The Glenlivet

REGION
Speyside

ADDRESS
Minmore, Ballindalloch, Moray

PHONE
01340 821720

WEBSITE
www.theglenlivet.com

OWNER
Chivas Brothers

VISITORS
Large visitor centre, with museum, restaurant and shop

CAPACITY
10.5m L.P.A.

HISTORICAL NOTES: George Smith of Upper Drummin farm, on the Duke of Gordon's Glenlivet Estate, was the first in the district to apply for a licence under the 1823 Spirits Act. Prior to this, like his neighbours, he distilled illegally: 'about a hogshead a week [250 litres] in the year following Waterloo [1815]'. When he 'went legal', his former colleagues saw this as an act of betrayal, and threatened to burn down his distillery – 'and him at the heart of it'. The pistols he carried for several years to ward off any such attempt are to be seen in the visitor centre today.

By the mid 1820s his whisky had won a reputation beyond the Highlands, and Andrew Usher of Edinburgh became his agent. In 1840 George Smith was able to lease another farm at Delnabo (where he took over a small distillery called Cairngorm from the previous tenant, John Gordon) and ten years later he leased Minmore, where he built a brand new distillery in 1859, the core of the present operation.

By this time Usher's was offering the first recorded 'branded Scotch' – Usher's Old Vatted Glenlivet – which started as a blended malt and finished (after about 1860) as a blended whisky.

George Smith was succeeded by his son, John Gordon Smith, in 1871.

The fame of Glenlivet whisky was such that many distilleries, some of them over 20 miles from the place itself (giving rise to Glenlivet being 'the longest glen in Scotland'), began to use the appellation, until J.G. Smith obtained a court order granting him sole rights to the definite article 'The' (1884), while others might use Glenlivet as a suffix only. By 1950 around 27 distilleries were doing this.

J.G. Smith died in 1901 and was succeeded by his nephew, George Smith Grant, who was in turn succeeded by his son, Bill, and grandson, Russell. Bill Smith Grant is the man who put malt whisky on the map of America. As soon as Prohibition ended in 1933 he began to look for business

In 2014 The Glenlivet became the best-selling malt in the world, displacing Glenfiddich, which had held this accolade since 1963.

partners and shipped a few hundred cases; by 1939 shipments had increased ninefold, including stocks of 2-oz miniatures for the famous Pullman Railway Company and Blue Ribbon inter-city express trains.

In the post-war period it enjoyed cult status, although there was an acute shortage of mature stock. Some cases were reserved for the luxury transatlantic liners, SS *United States* and SS *America*.

In 1953 G. & J.G. Smith Ltd amalgamated with J. & J. Grant Ltd to form The Glenlivet & Glen Grant Distilleries Ltd, which again amalgamated in 1970 with Longmorn-Glenlivet and the blending house Hill Thomson & Company to form The Glenlivet Distilleries Ltd. In 1973, the distillery was expanded from four to six stills, and went over to direct firing by gas, rather than coal. A large and ugly dark-grains plant was built close by in 1975, which ruined whatever charm the distillery had retained. Chivas Brothers have at least tidied this up.

The Glenlivet Distilleries Ltd was purchased by the Seagram Company in 1978. The new owner installed a further two direct fired stills that year, but it was not until the mid 1980s that all eight were converted to indirect heating by steam coils and pans.

In 2001 Chivas Brothers (Pernod Ricard) acquired The Glenlivet Distillery along with most of Seagram's other whisky interests. In June 2010 a new distillery within the existing distillery was opened by The Prince of Wales, with six new stills, eight new washbacks and a new mash tun, increasing capacity by 75%.

In 2014 a planning application to expand the distillery substantially was approved. The proposal is to build two new distilleries on the site, each with seven pairs of stills. When completed, this will triple The Glenlivet's capacity to 30m L.P.A.

CURIOSITIES: Glenlivet was a wild and remote place in the old days. It bred a bold and self-reliant people who clung not only to Roman Catholicism, but to

When George IV visited Edinburgh in October 1822, on his famous 'jaunt', arranged by Walter Scott, he drank nothing but (illicit) Glenlivet whisky.

the tradition of whisky-making, long after private distilling was banned in the 1780s. By 1820 it was estimated that there were 200 illicit stills in the glen, and the whisky made there had the highest reputation of any in Scotland.

In August 2007 The Glenlivet launched its Smugglers' Trails, in conjunction with the Crown Estates (which owns the land hereabouts). Three way-marked walks of differing lengths into the surrounding countryside allow visitors to follow the trails once used by smugglers.

'Gie me the real Glenlivet, and I weel believe I could mak' drinking toddy oot o' sea-water. The human mind never tires o' Glenlivet, any mair than o' caller air. If a body could just find oot the exac' proper proportion and quantity that ought to be drunk every day, and keep to that, I verily trow that he might leeve forever, without dying at a', and that doctors and kirk-yards would go oot o' fashion,' – James Hogg, quoted by Christopher North, 1826.

John Smith's Delnabo Distillery closed in 1858, owing to poor water quality and consolidation at Minmore, but he continued to lease the estate for stalking and shooting – an indication of his prosperity! John Gordon, the original distiller here, also had a distillery at Croughly, near Tomintoul, the remains of which can still be seen.

In 2014 The Glenlivet toppled Glenfiddich as the best-selling malt whisky in the world, selling 1.08 million nine-litre cases to Glenfiddich's 1.05. Glenfiddich had been the leading single malt since 1963.

'What makes The Glenlivet so special?', a *Time* magazine journalist asked Bill Smith Grant in the 1950s. 'There's nothing secret about it,' he replied. 'It just comes out like that . . . I think it's 99 per cent the water and a certain fiddle-faddle in the manufacture.'

Seagram's greatly expanded the visitor facilities in 1996/97, with a multi-media facility, restaurant

The Glenlivet is truly 'The Malt that started it all', as claimed in its advertising.

etc. The designers, who had earlier done an excellent job at Strathisla Distillery (and elsewhere), were confronted with a considerable challenge, which brings to mind converting sows' ears into silk purses, but they have succeeded admirably.

The site now welcomes around 45,000 people a year.

RAW MATERIALS: Mineral-rich, hard process water from Josie's Well; cooling water from bore-holes in the hills behind the distillery; reduction water from Blairfindy Spring. Own floor maltings until 1966, now unpeated malt from Crisp of Port Gordon, on the Moray Firth.

PLANT: Glenlivet now has two stillhouses. In the original is the old mash semi-Lauter tun (currently mothballed), eight Oregon pine washbacks, four lamp-glass wash stills (15,000 litres charge) and four lamp-glass spirit stills (10,000 litres charge), all indirect fired, with shell-and-tube condensers. The new stillhouse has a full-Lauter mash tun (13 tonnes capacity), eight wooden washbacks and three pairs of stills of the same size and style as the original.

MATURATION: Ex-bourbon casks for the 12YO, ex-sherry casks for the 18YO; the rest are a mix of bourbon, sherry and refill casks.

STYLE: Medium-bodied, complex Speyside, with fruits (pineapple, pears, apples) and floral notes.

MATURE CHARACTER: The Glenlivet is a complex malt, and develops depth and extra complexity with age. The younger expressions are soft, floral and fruity (apple, pineapple and peach). Some malty notes are apparent in both aroma and taste, and some vanilla from the American oak. The taste starts sweet, with a soft texture, and dries considerably in the finish. Medium-bodied.

Glenlochy

REGION
Highland (West)
ADDRESS
North Road,
Inverlochy,
Fort William,
Inverness-shire
LAST OWNER
D.C.L./S.M.D.
CLOSED
1983

As well as buying distilleries for N.D.A. and for himself, in 1939 Joe Hobbs wisely bought a company in Norfolk which made fire extinguishers.

HISTORICAL NOTES: April 1900 was not the best time to open a distillery. Pattison's of Leith had just gone bust, bringing down a number of other companies; confidence in the industry had collapsed and maturing stocks were far greater than the market could cope with.

Three years earlier, when David McAndie bought the ground from Lord Abinger of Inverlochy Estate, the picture was brighter. Ben Nevis, Fort William's first distillery, was thriving and the West Highland Railway had reached the town in 1894. McAndie, who had recently built Glen Cawdor Distillery, Nairn, was joined by James Grant, owner of Highland Park Distillery, and 13 local investors. The first manager came from Balblair Distillery.

In spite of an unpropitious start, Glenlochy remained in production until 1917, when all distilleries were closed by government order. In 1920 the shareholders sold to a consortium of Lancashire brewers; they resumed production in 1924 and closed again after two years. In 1934 the buildings and site were sold for £850 to a Lancashire car hirer, who sold them on three years later to Joseph Hobbs, along with the Glenlochy Distillery Company (see *Ben Nevis*).

Hobbs had a remit to buy distilleries from National Distillers of America, Inc., and sold Glenlochy to them in 1940 to raise money for his 'Great Glen Cattle Ranch'. D.C.L. took over National Distillers in 1953 and transferred management at Glenlochy to S.M.D., which modernised it in 1960 and 1976, then closed the distillery in 1983.

CURIOSITIES: The exterior of the distillery was built entirely of brick (unusually) and is decorated on the roof with ironwork; the pagoda has an unusually steep pitch. It has changed little.

Joseph Hobbs was a colourful character. His parents emigrated to Canada when he was a child; as an adult he made a fortune in shipbuilding and property development (and in running whisky during Prohibition), then sustained heavy losses during the Great Depression of 1930/31, and returned to England with less than £1,000, according to his obituary in *The Times*.

G1 Glenlossie

REGION
Speyside

ADDRESS
Thomshill, by Elgin,
Moray

PHONE
01343 862000

WEBSITE
www.malts.com

OWNER
Diageo plc

VISITORS
By appointment

CAPACITY
3.7m L.P.A.

HISTORICAL NOTES: Glenlossie Distillery was built by a publican and a few friends in 1876: John Duff, tenant of the Fife Arms at Llanbryde and former Manager of Glendronach Distillery, in partnership with Alexander Grigor Allen (Procurator Fiscal of Morayshire, and after 1880 part owner of Talisker Distillery), H.M.S. Mackay (land agent and burgh surveyor of Elgin) and John Hopkins (a London-based blender, and owner from 1880 of Tobermory Distillery and of a well-known blend, Old Mull). Hopkins was agent for the new distillery's make. In 1896, when the Glenlossie-Glenlivet Distillery Company was founded, Mackay took over management. The company joined D.C.L. in 1919 and was managed by S.M.D. (licensed to John Haig & Company) from 1930.

In 1962 Glenlossie was expanded from four to six stills, and ten years later a new distillery named Mannochmore was built adjacent (see entry), to operate in parallel with the original – part of S.M.D.'s expansion plans in the early 1960s (see *Caol Ila* etc.). A large dark-grains plant was built on the site (1968–72), with a conspicuous white chimney which can be seen for miles around. Each week it can process 2,600 tonnes of draff and eight million litres of pot ale from 21 distilleries, to produce 1,000 tonnes of cattle feed.

CURIOSITIES: 'With the exception of the Distilling House (which is built of stone) the distillery is constructed entirely of cement which, under the sunlight as we descended the hill, looked beautifully white and clean.' (Alfred Barnard, 1887)

A private railway siding on the Perth–Elgin line gave direct access to southern markets after 1896. 'Glenlossie-Glenlivet was efficiently managed … New Warehouses were built, and extension or improvements effected in almost every year up to 1917, when all malt whisky distilleries were closed.' (Brian Spiller)

A horse-drawn fire engine, still preserved at Glenlossie, was little help in quelling a fire that destroyed part of the distillery in 1929.

RAW MATERIALS: Process water from the Bardon Burn, which sources in the Mannoch Hills; cooling water from the Gedloch Burn and the Burn of Foths. Floor maltings until 1962; now unpeated malt from Burghead.

PLANT: An unusual semi-Lauter mash tun, combining aspects of a Steiniker tun with the original Newmill Lauter knives (eight tonnes) – ensuring high efficiency. Eight larch washbacks. Three plain wash stills (15,500 litres charge); three plain spirit stills (17,500 litres charge); all with purifiers. All indirect fired. Shell-and-tube condensers.

MATURATION: On site in traditional and racked warehouses, and in the Central Belt.

STYLE: Grassy.

MATURE CHARACTER: A fresh, light, Speyside nose, with cut grass, faint flowers and hair lacquer. The taste is sweet and perfumed. Light-bodied.

Glen Mhor Demolished

REGION
Highland (North)
ADDRESS
Telford Street, Muirtown,
Inverness, Inverness-shire
LAST OWNER
D.C.L. /S.M.D.
CLOSED
1983

Glen Mhor has been described as a classic example of 'The Highland Style'.

HISTORICAL NOTES: John Birnie, Manager of Glen Albyn Distillery and largely responsible for its success (output had trebled between 1887 and 1892), was frustrated in his ambition to buy shares in the distillery, so he formed a partnership with Charles Mackinlay & Company, wine and whisky merchants in Leith, bought the site across the road from Glen Albyn and alongside the Caledonian Canal, and built Glen Mhor Distillery. It went into production in 1894, designed by Charles Doig and with power supplied by a 30-foot-high turbine wheel driven by water from the canal. Electric motors only replaced it in the early 1950s, and even then it continued to be used for driving the mills and the washbacks until about 1960.

Mackinlay & Birnie became a limited company in 1906, with John Walker & Sons holding 40% of the shares, and bought Glen Albyn in 1920. A third still was added at Glen Mhor by 1925, and all stills went over to internal heating by steam in 1963. A Saladin box maltings was installed in 1949. Mackinlay & Birnie was acquired by D.C.L. in 1972 and transferred to S.M.D.

Glen Mhor was demolished in 1983 and the site is now a shopping centre.

CURIOSITIES: Supplies to both Glen Mhor and Glen Albyn were delivered by sea via the Caledonian Canal, including peat for the malt kiln, which came from Orkney.

Harper's Weekly, the trade gazette, commented: 'There is nothing decorative or grand about the appearance of the buildings; nevertheless they are most solidly constructed and well arranged for carrying on the work efficiently.'

Glenmorangie

REGION
Highland (North)
ADDRESS
Tain, Ross-shire
PHONE
01862 892477
WEBSITE
www.glenmorangie.com
OWNER
Glenmorangie plc (LVMH
Moët Hennessy Louis
Vuitton SE)
VISITORS
Visitor centre, with
museum (opened 1997)
and shop
CAPACITY
6m L.P.A.

There have been only seven distillery managers at Glenmorangie since its foundation.

HISTORICAL NOTES: Glenmorangie has been the best-selling single malt in Scotland since 1982, standing number two in the U.K. and number four in world sales. Creative advertising in the early 1980s may well have helped.

The distillery stands near the ancient Royal Burgh of Tain, in Ross-shire, overlooking the Dornoch Firth, on Morangie Farm, a spot well-known to illicit distillers. It was established in 1843 by William and John Matheson. The former was part-owner of Balblair Distillery and related to Alexander Matheson who founded Dalmore (see entry). By 1849 production had reached 20,000 gallons. A limited company was formed in 1887, and the distillery completely rebuilt the same year.

In 1918 it was sold to Macdonald & Muir (40% share), the established Leith firm of blenders, and Durham & Company, a whisky broker (60%). By the late 1930s, M. & M. had acquired Durham's share and the malt was being used in their blends (notably Highland Queen and Martins V.V.O.); although it could be bought in bulk, Glenmorangie was not promoted as a single until the late 1970s. In 1979 capacity was doubled (to four stills), then doubled again in 1990.

By the mid 1980s, M. & M. was experimenting with different wood types and with re-racking into wine barrels, in order to increase the range of expressions. The light style of Glenmorangie spirit does not lend itself to complete maturation in European oak, but can benefit from 'finishing' in such casks. A sherry-finished 18YO (introduced 1992) was soon followed by a port-wood and a Madeira-wood; these were followed by more than a dozen other styles.

In 1996 Glenmorangie became a public limited company; in 2004 it bought the Scotch Malt Whisky Society, the Macdonald family sold its shares, and both companies were acquired by the French luxury goods giant, LVMH Moët Hennessy Louis Vuitton SE.

So keen is Glenmorangie on its wood that the company owns forests in the Ozark Mountains of Kentucky.

In October 2008, capacity was increased to six million L.P.A., by the addition of four new stills, four new washbacks and a new mash tun.

CURIOSITIES: Glenmorangie was one of the first distilleries to heat its wash still indirectly, by steam coil (1883). The stills themselves, which are the tallest in Scotland – 'as tall as a giraffe' –, were originally used for distilling gin and the elegant shape has been retained.

Although there are records of Glenmorangie being shipped overseas (to the Vatican and San Francisco) in 1880, it was only in the late 1970s that its owners began to focus on their malt. In 1981 the 'Sixteen Men of Tain' print campaign was launched. This emphasised the high craft that went into making the malt, while humanising it – the ads featured wood-cuts of the men themselves, from Distillery Manager to tractorman. Ahead of its time, it ran for 20 years.

Both distilleries were acquired by LVMH Moët Hennessy Louis Vuitton SE in 2004 and in 2007 Glenmorangie announced they were repackaging and renaming their key products. The familiar tan label with black and red lettering and a chequered border, introduced in the 1970s, has been replaced by cleaner, brighter, more sophisticated labelling, which nevertheless retains elements from the past. Bottle shapes have become more sensual, and the overall impression is more feminine.

Not far from the distillery, Glenmorangie owns a magnificently restored mansion, now named Glenmorangie House. Close to it is an important Pictish carved stone, the Cadboll Stone. The new packaging draws upon Celtic symbols from the stone to good effect.

RAW MATERIALS: The Tarlogie Spring – very hard, mineral-rich water. Floor maltings until 1977; now unpeated malt from independent maltsters.

Glenmorangie's first stills came from a gin distillery, and are the tallest in Scotland. They worked so well that the shape and size have been copied ever since.

PLANT: Stainless steel full-Lauter tun (9.8 tonnes per mash), six stainless steel washbacks. Six very tall (5.18m) boil-pot wash stills (11,300 litres charge); six very tall boil-pot spirit stills (7,500 litres charge). All indirect fired by steam. Shell-and-tube condensers.

MATURATION: First- and second-fill ex-bourbon American white oak casks; some other woods for finishing. Fourteen bonded warehouses on site; ten traditional dunnage, four racked to eleven casks high.

STYLE: Light, floral, citric (tangerine).

MATURE CHARACTER: Glenmorangie is light in style, but complex in character. Keynotes are vanilla, almonds, mandarins, apples, roses, spices and hay. The mouthfeel is soft and fresh, the taste lightly sweet, with some cereal and fresh fruit. Medium-bodied.

Glen Moray

REGION
Speyside

ADDRESS
Bruceland Road, Elgin, Moray

PHONE
01343 542577

WEBSITE
www.glenmoray.com

OWNER
La Martiniquaise

VISITORS
Visitor centre, shop, café, tasting area

CAPACITY
3.3m L.P.A.

HISTORICAL NOTES: Like several distilleries, Glen Moray was built on the site of an early nineteenth-century brewery, to take advantage of a reliable water source. Unlike any other distillery, however, the grounds included the site of Elgin's gallows, on the edge of the town, beside the main road heading west; a grim warning to unruly Highlanders arriving in the city.

The distillery was built by a local consortium, the Glen Moray-Glenlivet Distillery Company Ltd, in 1897, making use of some of the old West Brewery buildings and 'being equipped with electric light throughout'. By 1910 it had closed, and it was acquired by Macdonald & Muir (see *Glenmorangie*) in 1920, resuming production in 1923. They extended from two to four stills in 1958, making use of the make in their successful blends, notably Highland Queen.

Glen Moray single malt began to be released in 1976, but was only promoted from the early 1990s. Macdonald & Muir began their experiments with wood-finishing at Glen Moray – now a familiar feature of the Glenmorangie portfolio. Three wood-finished expressions were released in 1999: Chardonnay (no age), and Chenin Blanc (12 and 16YO).

Both distilleries were acquired by Louis Vuitton Moët Hennessy in 2004, and in September 2008 L.V.M.H. sold Glen Moray to the French spirits company, La Martiniquaise. In 2012/13 the new owner installed two more stills, in the same style as the existing stills, increasing capacity by over a third.

CURIOSITIES: Glen Moray has had only five managers in its 110-year lifetime. The last to retire (after 18 years, in 2005) was Ed Dodson, in whose honour a Manager's Choice from 1962 was released.

RAW MATERIALS: Hard water from the River Lossie. Floor maltings until 1958 then Saladin box maltings until 1977; now unpeated malt from independent maltsters.

PLANT: Stainless steel semi-Lauter tun with peaked canopy (7.5 tonnes). Five stainless steel washbacks. Three plain wash stills (10,000 litres charge); three plain spirit stills (6,000 litres charge). Indirect fired. Shell-and-tube condensers.

MATURATION: Mainly first and refill ex-bourbon casks. Small amount of ex-sherry casks. Mix of dunnage and palletised warehouses on site, holding 65,000 casks in total.

STYLE: Fruity, floral, clean.

MATURE CHARACTER: Glen Moray is light in style, in its younger expressions, but elegantly balanced and 'well made'. The nose is floral and fruity, with butterscotch, vanilla and barley sugar. The taste is sweet, with nutty notes and a delicate fruitiness. Medium-bodied.

Glen Ord

REGION
Highland (North)

ADDRESS
Muir of Ord, Ross and Cromarty

PHONE:
01463 872004

WEBSITE
www.malts.com

OWNER
Diageo plc

VISITORS
Spacious visitor centre, with interesting artefacts

CAPACITY
11m L.P.A.

HISTORICAL NOTES: Like many others, the distillery at Muir of Ord was established on a site which had formerly been much used by smugglers (over 40 illicit stills), and which was famous for whisky production, being near Ferintosh and the Black Isle. The museum displays several illicit stills dredged from lochs nearby.

In addition to this illicit activity, there were nine licensed distilleries in the district, all except one worked by farmers' co-operatives. The *New Statistical Account* (1840) states that 'distilling is the sole manufacture of the district'.

Ord Distillery was founded in 1838 by Thomas Mackenzie, the land owner, and licensed to Robert Johnstone and Donald McLennan. Johnstone went bankrupt in 1843, as did his successor, Alexander McLennan. When he died in 1870 it passed to his widow, who prudently married Alex MacKenzie, a banker. In 1878 he built a new stillhouse, but this burned down the same year, so he started again. He died in 1896 and ownership passed to James Watson & Company of Dundee, who extended the distillery and tripled production by 1901. When the last surviving Watson, John Jabez, died in 1923 Ord was sold to John Dewar & Sons, and thus passed to D.C.L. in 1925, and to S.M.D. in 1930.

Floor maltings were changed to Saladin boxes in 1961, and a large drum maltings built on an adjacent site in 1968. The distillery itself was rebuilt and expanded in 1966, in the 'Waterloo Street' design (see *Caol Ila, Glendullan* etc.), masterminded by Dr Charlie Potts, S.M.D.'s Chief Engineer. The visitor centre was opened in 1988, and today welcomes around 20,000 visitors a year.

In 2010/11 a new mash tun was installed and two more washbacks; then in 2014/15 Diageo invested £25 million to double Glen Ord's capacity. Eight new stills are located in the distillery's former Saladin maltings, with 12 wooden washbacks in the former kiln and malt storage area. Capacity is now

Alfred Barnard, who visited in the mid 1880s, reported that whisky was being shipped to 'Singapore, South Africa and other colonies' under the brand name Glen Oran. He 'tasted some 1882 make and found it very agreeable to the palate'.

11 million L.P.A., placing Glen Ord among the top five largest malt distilleries in Scotland.

CURIOSITIES: Glen Ord has its own sizeable maltings (Saladin boxes 1961–83, 18 drums added 1968), producing its own requirement and that of the seven other distilleries in the north of Scotland owned by S.M.D. at that time. Over the years the whisky has been branded as Glenordie, Ord, Glen Oran, Ordie and Muir of Ord.

The distillery was lit by paraffin lamps until 1949, with a water turbine as the main source of power (until 1961).

Ord was the main S.M.D. site for experiments in heating stills. Until 1958 all four stills were direct fired by coal; two were converted to direct oil firing that year, and in 1962 to indirect heating by steam. Each time the distillates were compared for consistency. In 1966, the two direct fired stills were converted, and two more stills installed. Hot water from the condensers is directed to the maltings, so they are run unusually hot, and condensation is completed by horizontally mounted 'after-coolers'.

The Singleton of Glen Ord was launched in Asia in 2006: by 2013 it was selling over two million bottles a year, and was numbered among the top ten best-sellers worldwide. Yet it still accounts for only around 15% of Glen Ord's output, so it is not surprising that the distillery is being substantially expanded.

RAW MATERIALS: Lightly peated malt from own maltings. 85% of barley grown locally. Water via Alt Fionnadh (the White Burn) from two lochs (the 'Loch of the Peats', *Loch nam Bonnach*, and the 'Loch of the Birds', *Loch nan Eun*).

PLANT: Large cast iron semi-Lauter mash tun with copper dome (12.5 tonnes); eight Oregon pine washbacks. Three plain wash stills (18,000 litres charge); three plain spirit stills (16,000 litres charge). All steam

Excavations here in 1962, to make way for No. 1 warehouse, uncovered six skulls, one with a musket ball embedded in its jaw.

fired. Shell-and-tube condensers, and after-coolers on all stills.

MATURATION: Approximately half first and refill sherry butts, and half refill hogsheads, for single malt bottlings. 12,500 casks matured on site in dunnage warehouse, the rest tankered to Cambus or to customers' own sites.

STYLE: Described as 'the best representative of the "Highland" style'. Sweet and heathery, with distinct waxiness.

MATURE CHARACTER: 'The Singleton of Glen Ord' retains the malt's characteristic waxy texture and light smokiness, while adding depth and complexity. A big, rich nose with fruity (nectarines and dried orange peel) and floral (old-fashioned perfume) notes, flaked almonds and sandalwood. Very smooth and chewy texture; sweet, slightly mouth-cooling, with a long finish and a pleasant aftertaste of nougat. Medium-bodied.

Glenrothes

REGION
Speyside

ADDRESS
Burnside Street, Rothes,
Moray

PHONE
01340 872300

WEBSITE
www.theglenrothes.com

OWNER
The Edrington Group Ltd

VISITORS
Trade only

CAPACITY
5.6m L.P.A.

HISTORICAL NOTES: In 1868, James Stuart, a 'corn factor' in Rothes village on Speyside and proprietor of the Mills of Rothes, took over Macallan Distillery nearby, in partnership with Robert Dick and William Grant (agents for the Caledonian Bank), and John Cruikshank (solicitor). Macallan prospered and three years later they decided to build a second and larger distillery upstream from Mills of Rothes.

This was not a propitious time to be building a distillery; by midsummer 1878 Britain was involved in 'the worst economic crisis for over a century'; in October the City of Glasgow Bank collapsed and in December the Caledonian Bank closed its doors. James Stuart & Partners was dissolved, Stuart remaining at Macallan and the others taking over the half-built Glen Rothes-Glenlivet Distillery, trading as William Grant & Company. It opened on 28 December 1879, the night of the Tay Bridge Disaster (and see *Balmenach*), with Robertson & Baxter appointed as agents. On the suggestion of W.A. Robertson, Glen Rothes was merged with Bunnahabhain Distillery in 1887 to form the Highland Distilleries Company.

In 1896 the distillery was expanded to designs by Charles Doig of Elgin, with four stills and a second malt kiln, but a fire destroyed much of the distillery before the work was completed. In spite of installing Doig's patent appliances for preventing explosions in the new mill-room, this failed to avert another fire six years later.

Glen Rothes was enlarged to six stills in 1963 and to eight in 1980. In 1989 it was rebuilt with ten stills. In 1987 Highland licensed the The Glenrothes brand to Berry Brothers & Rudd, the old established London wine merchants, which also owned 50% of Cutty Sark (the other 50% being owned by

Robertson & Baxter), and they released the first 'official' bottling that year, at 12 years old.

In 2010 Edrington (owner of Robertson & Baxter) took 100% ownership of Cutty Sark and Berry Brothers 100% ownership of the The Glenrothes brand.

Until 1994 the distillery was named 'Glen Rothes'; since then it has been 'Glenrothes'.

CURIOSITIES: Twenty-seven Speyside distilleries adopted the suffix 'Glenlivet', owing to the fame of the original. In 1884 John Gordon Smith raised an action to limit this usage, but before it came to court the owners of Glen Rothes, Cragganmore, Mortlach, Glenfarclas, Linkwood, Glengrant (sic), Glenlossie and Benrinnes agreed to Smith's distillery being The Glenlivet, while they were 'permitted to add Glenlivet to their name and sell blends of their whiskies as Blended Glenlivet'. The remainder followed suit. Andrew Usher was allowed to continue selling Old Vatted Glenlivet.

The Lady's Well, from which the distillery draws its process water, was the site of a murder in the thirteenth century. Mary Leslie, daughter of the Earl of Rothes, was killed by the notorious Wolf of Badenoch (Alexander Stewart, Earl of Buchan, and son of the King of Scots) while trying to protect her lover.

As well as relabelling in 1994, Glenrothes adopted a unique dumpy bottle shape, known affectionately as La Bomba. In 2005 they introduced La Bombette, a hand-grenade-sized 10cl mini bottle, which holds two large measures.

RAW MATERIALS: Water from the Lady's Well and Ardcanny Spring. Lightly peated malt from Tamdhu Distillery maltings and from Simpson's.

A fire in 1922 engulfed No. 1 warehouse, pitching large amounts of mature whisky into the Rothes Burn, to the pleasure of locals and, it is said, even passing cows.

PLANT: Stainless steel semi-Lauter mash tun (4.92 tonnes). Twelve Oregon pine washbacks and eight stainless steel washbacks (latter not currently in use). Five boil-pot wash stills (12,750 litres charge); five boil-pot spirit stills (15,000 litres charge), all indirect fired by steam. Shell-and-tube condensers.

MATURATION: American oak casks (ex-sherry and ex-bourbon), European oak ex-sherry. Four racked and 12 dunnage warehouses are on site.

STYLE: Sweet and fruity, but also heavy and rich.

MATURE CHARACTER: Glenrothes is a heavy style of malt, and takes sherry-wood maturation well, and advanced age. The Select Reserve best displays the distillery character: on the nose, nougat, dried fruits and nuts, caramel, vanilla sponge. Sweet to taste, with a soft texture, drying towards the end, with a nutty aftertaste and hints of chocolate. A useful combination of freshness and depth. Medium-bodied.

Glen Scotia

REGION
Campbeltown

ADDRESS
High Street,
Campbeltown, Argyll

PHONE
01586 552288

WEBSITE
www.glenscotia.com

OWNER
Loch Lomond Group Ltd

VISITORS
Visitor centre, with
tastings and shop

CAPACITY
600,000 L P A

HISTORICAL NOTES: Glen Scotia has had a patchy existence, but, unlike 30 other distilleries that once operated in Campbeltown, it has survived.

The distillery was established in 1832, named simply Scotia, by Stewart, Galbraith & Company, which became a limited company in 1895 and sold to West Highland Malt Distilleries in 1919. Duncan MacCallum, a well-known local distiller, bought Scotia in 1924, just before the Great Depression. It was closed from 1928 to 1930, then sold to Bloch Brothers after MacCallum's suicide in 1930; Sir Maurice Bloch sold it and Scapa Distillery to Hiram Walker in 1954.

A Gillies & Company bought Glen Scotia the following year and operated it until 1984. They sold to Gibson International in 1989 (owner of Littlemill Distillery), which went into receivership in 1994, when both Glen Scotia and Littlemill were acquired by A. Bulloch & Co., who also owned Glen Catrine Bonded Warehouse, in Mauchline (established in 1974 and now one of the largest bottling plants in Scotland), and Loch Lomond Distillery (which Sandy Bulloch had bought in 1986). Glen Scotia was mothballed, then operated for only a couple of weeks a year by a team from Springbank Distillery. In 2007 it went into part-time production (producing around 100,000 L.P.A. per annum).

In March 2014, after over a year's negotiation, the holding company for all these enterprises, Loch Lomond Distillers Ltd, was bought by a group of senior managers with the support of a private equity company. The new owner, named 'Loch Lomond Group Ltd', was incorporated specifically for the purpose, under the leadership of Colin Matthews, formerly a senior executive of Imperial Tobacco, and Nick Rose, formerly C.F.O. at Diageo, with Exponent Private Equity as its

MacCallum's ghost has been seen at Glen Scotia, and his death supposedly inspired the well-known song which starts 'Oh, Campbeltown Loch, I wish ye were whisky'.

holding company. The distillery has been thoroughly upgraded and expanded, with new washbacks, new roofs, new spirit safe and a visitor centre. A new range of expressions was released in early 2015: Double Cask (ex-bourbon, finished in PX), 15 years old and Victoriana (cask strength).

CURIOSITIES: As David Stirk writes in his book *The Distilleries of Campbeltown*, 'the nadir of [the decline in Campbeltown distilling] is perhaps most poignantly represented with the suicide on 23 December 1930 of Duncan MacCallum, aged 83, once a leading distiller in the town, when he drowned himself in Crosshill Loch'.

Bloch Brothers changed the name from Scotia to Glen Scotia in 1934/35. In a press release dated October 1940, they state that much of the make is exported to America 'after blending'.

RAW MATERIALS: Soft water from Crosshill Loch. Unpeated and lightly peated (four to five weeks per annum) from Greencore maltings, Buckie.

PLANT: Corten steel rake-and-plough mash tun with a canopy (2.72 and 1.92 tonnes per mash at the moment). Six Corten steel washbacks. One plain wash still (7,500 litres charge), one plain spirit still (8,400 litres charge), both indirect fired. Shell-and-tube condensers.

MATURATION: Bourbon barrels mostly, with some sherry hogsheads and butts. Single racked warehouse on site holding 6,500 casks, all matured on site.

STYLE: Maritime and oily.

MATURE CHARACTER: Glen Scotia has a 'maritime character': seaweed, docks, briny, lightly peaty. The taste is somewhat oily and smooth, with cereal notes, nuts, light sweetness, distinct saltiness and some shoreline seaweed notes. Dry overall and medium-bodied. Variable.

Glen Spey

REGION
Speyside

ADDRESS
Rothes, Moray

PHONE
01340 831215

WEBSITE
www.malts.com

OWNER
Diageo plc

VISITOR
By appointment

CAPACITY
1.4m L.P.A.

HISTORICAL NOTES: James Stuart, grain merchant in Rothes, licensee (1868–86) then owner (1886–92) of Macallan Distillery (see also *Glenrothes* Distillery), built Glen Spey (originally a corn mill, and named Mill of Rothes Distillery). He sold it the year after he bought Macallan to W. & A. Gilbey, the London wine and spirits merchant, for £11,000. This was the first time an English company had bought a Scotch whisky distillery. They went on to build Knockando Distillery in 1904.

In 1962, Gilbey's merged with Justerini & Brooks, another long established London wine and spirits merchant, enjoying considerable success in America with its blend, J. & B. Rare. The new company was named Independent Distillers & Vintners (I.D.V.); Glen Spey was doubled in capacity in 1970, sold to Watney Mann (brewers) two years later, and acquired by Grand Metropolitan the same year. Grand Metropolitan merged with Guinness in 1997 to form U.D.V., now called Diageo.

Glen Spey has long been a key filling for the J. & B. blends. For a time it was the own-label malt in Unwin's off-licences, but it was not bottled as a single by its owners until 2001.

CURIOSITIES: Glen Spey is discreetly tucked in off the main street in Rothes, below a fragment of the outer wall of the once formidable Rothes Castle. The castle dates from the twelfth century, and King Edward I of England lodged there in 1296. By 1309 it was owned by the powerful Leslie family, created Earls of 'Rothays' in the sixteenth century, and Dukes of Rothes in 1680 (this only lasted one year). The castle and adjoining buildings were burned by the locals 'to prevent thieves from harbouring in it' in 1662.

In the days of direct firing by coal and lighting by naked flames, distilleries were plagued by fire (see *Talisker*, *Glenrothes*, *Glenlossie* etc.). Glen Spey was badly damaged by a fire in 1920.

During World War II troops were billeted at Glen Spey. One soldier was electrocuted, and it is said that his ghost still haunts the distillery.

RAW MATERIALS: Floor maltings until 1969, now unpeated malt from Burghead. Soft process water from the Doonie Spring; cooling water from the River Rothes.

PLANT: Semi-Lauter mash tun (4.4 tonnes). Eight stainless steel washbacks. Two lamp-glass wash stills (10,600 litres charge); two lamp-glass spirit stills (7,000 litres charge), with purifiers. All indirect fired. Shell-and-tube condensers.

MATURATION: Refill ex-bourbon hogsheads. Mainly in the Central Belt.

STYLE: Nutty-spicy, light.

MATURE CHARACTER: A light style of malt, made for blending. The nutty, cereal character comes through on the nose, with a Speyside floral element. The taste is sweet, with a malty, grassy character. Some have identified roast chestnuts. Shortish finish. Innocuous. Medium- to light-bodied, all by design.

Glentauchers

REGION
Speyside

ADDRESS
Mulben, Keith, Banffshire

PHONE
01542 860272

WEBSITE
No

OWNER
Chivas Brothers

VISITORS
No

CAPACITY
4.2m L.P.A.

HISTORICAL NOTES: Glentauchers was founded in 1897 by James Buchanan & Company (of Black & White fame) in partnership with their spirit supplier, W.P. Lowrie: the former bought out the latter in 1906. The attractive buildings three miles outside Keith were designed by John Alcock, under the supervision of the legendary Charles Doig. Like many other distilleries of the period, the site was chosen for its water supply and communications: a railway siding was constructed at the back of the site. It went into production in May 1898.

Curiously, experiments in the continuous distillation of malt whisky were conducted here around 1910. The plans for the 'continuous pot still' form part of the Doig Collection in Moray District Library, Elgin. There is also an example from Convalmore Distillery in the Whisky Museum, Dufftown.

Buchanan's joined D.C.L. in the Big Amalgamation of 1925, and Glentauchers was managed by S.M.D. from 1930, but licensed to James Buchanan & Company until 1949.

During the general refurbishment of distilleries in the mid 1960s, Glentauchers was expanded from two to six stills (1966), with a new stillhouse built alongside its maltings. These were closed in 1968, although the two pagoda-topped Doig kilns remain and the building still houses a single drum malting dating from about 1925.

By the mid 1980s the pendulum had swung the other way. Glentauchers was mothballed in 1985, and then sold to Allied Distillers in 1989. Production recommenced in August that year.

Ownership changed again in 2005, when Pernod Ricard bought most of Allied's Scotch whisky interests. Cast iron tanks in the stillhouse were replaced in 2006 (some of them dated from the 1930s), although the wooden Intermediate Spirit Receiver was retained. The mashhouse was upgraded in 2007. At that time it was determined

'The extraordinary
thing is that the
possibility of failure
never once occurred
to me.

I had it always
before me in my
mind that sooner or
later I was bound to
make a success.'
James Buchanan

that Glentauchers would remain a manual operation (unlike most of Chivas Brothers' other distilleries), to provide a practical training facility for all staff, including management.

Glentauchers has increased capacity from 3.4 to 4.2 million L.P.A. since 2011, by moving to seven-day production and two washbacks from the demolished Caperdonich Distillery.

CURIOSITIES: James Buchanan went to London as an agent for Chas MacKinlay & Company in 1879, aged 30.

By the time of his death, aged 86, Buchanan was a peer (Lord Woolavington), owned estates in four counties, had twice won the Derby and had given enormous sums to charitable causes and hospitals. He left over seven million pounds – a huge fortune in 1935.

RAW MATERIALS: Production water from two reservoirs fed by the Rosarie and Tauchers Burns; the former supplies process water and the latter cooling water. Maltings closed 1968, now unpeated malt from independent maltsters.

PLANT: Full-Lauter mash tun (12 tonnes) since 2007. Eight larch washbacks. Three plain wash stills (10,000 litres charge), three plain spirit stills (6,300 litres charge). Indirect fired by steam pans and coils. Shell-and-tube condensers.

MATURATION: Refill American oak casks.

STYLE: Medium-bodied Speyside – sweet and fruity.

MATURE CHARACTER: Another malt which is principally made for blending. Broadly Speyside in style – sweet, fragrant, fruity, estery – Glentauchers adds a dash of coconut, almonds and cereal to the profile. Sweet to taste, with little body but a pleasant, summery impression. Light-bodied.

Glenturret

REGION
Highland (South)
ADDRESS
The Hosh, Crieff,
Perthshire
PHONE
01764 656565
WEBSITE
www.theglenturret.com
OWNER
The Edrington Group Ltd
VISITORS
Popular visitor centre
for all the family
The Famous Grouse
Experience – with
restaurant, shop, tours
and tastings
CAPACITY
340,000 L.P.A.

HISTORICAL NOTES: There was an illicit farm distillery at The Hosh from 1775, and based on this, Glenturret claims to be the oldest distillery in Scotland. The first licence was granted to John Drummond in 1818. He continued until he went bankrupt in 1842. John McCallum followed him (1852–74) but also went bust; then it was taken over by Thomas Stewart, who changed its name from 'Hosh' to 'Glenturret' in 1875. (There had been another Glenturret Distillery nearby, but this folded in the 1850s.)

Again it changed hands in 1903, to those of Mitchell Brothers Ltd; production ceased in 1921, and in 1929 Mitchell's went into liquidation and the distillery was dismantled.

Glenturret's revival was owing to James Fairlie, who bought the site in 1957 and reinstated the equipment 1959–60 – often using second-hand plant and integrating new buildings with the old – with a view to 'preserving the craft traditions of malt whisky distilling and developing its appreciation'. To the latter end, he opened to the public and arranged tours and tastings – if Glenfiddich was the first distillery to do this (in 1969), Glenturret was a close second. The British prime minister, Sir Alec Douglas-Home, visited in 1964.

In 1981, James Fairlie sold the distillery to Rémy Cointreau, who greatly expanded the visitor facilities. At that time, Rémy had a trading relationship with Highland Distilleries, and in 1993 Glenturret joined Highland Distilleries (now Edrington). In 2002 Edrington invested £2.2 million in further upgrading the visitor facilities, now named 'The Famous Grouse Experience', although the distillery had no previous connection to The Famous Grouse.

CURIOSITIES: Illicit distilling may have been carried on at The Hosh from before 1775: parish records reveal there were once many stills in the district, going back to 1717. Maybe this was due to the two hills which encompass it providing good vantage

Glenturret was long famous for its cat, Towser, who is commemorated by a bronze statue. She died in 1987, aged 24, having dispatched 28,899 mice – a fact which is recorded in The Guinness Book of Records!

points from which to observe hostile militia or excisemen.

Alfred Barnard found Glenturret to be antiquated in the 1880s; it remains so today, by design. The Famous Grouse Experience welcomed 90,000 visitors in 2013 and could claim to be the 'most visited distillery' that year. Its proximity to the popular spa town of Crieff helps.

Glenturret was first bottled as a single as early as 1965, but may only have been locally available.

RAW MATERIALS: Soft water from Loch Turret. Lightly peated malt from Simpson's.

PLANT: Open stainless mash tun (one tonne per mash), stirred by hand with a wooden paddle (this is unique). Eight Douglas fir washbacks (unusually, two washbacks charge the wash still). One boil-ball wash still (12,600 litres charge), one plain spirit still (6,800 litres charge; pre 1972, the spirit still was the wash still). Both indirect fired by steam. Shell-and-tube condensers.

MATURATION: Refill American and Spanish oak, six bonded warehouses on site. Some goes to Buckley Bond, Bishopbriggs.

STYLE: Fruity (orange), floral, slightly medicinal, with cereal notes.

MATURE CHARACTER: The younger expressions retain the distillery character well: floral, nutty, malty, with a hint of smoke. The taste is sweetish, with honey traces, nuts and cereal, and a thread of smoke. Medium-bodied.

Glenugie Demolished

REGION
Highland (East)
ADDRESS
Invernettie, by Peterhead,
Aberdeenshire
LAST OWNER
Whitbread & Company
Ltd
CLOSED
1983

Glenugie takes its name from the river that meets the sea at Inverugie, on the northern edge of Peterhead, the leading fishing port in the 1980s.

HISTORICAL NOTES: The distillery was established in the early 1830s on the site of an old windmill (the stump of which still survives) and was named Invernettie until 1837, when it was converted into a brewery. It was converted back in 1875 by the Scottish Highland Distillers Company Ltd, which was wound up only six years later. It passed through several hands, with periods of silence, including from 1925 to 1937.

In 1937 it was bought by Seager Evans, gin distillers in London, who had built Strathclyde Grain Spirit Distillery in 1927 (see entry) and acquired the blended whisky Long John in 1936, Seager Evans sold out to Schenley Industries Inc., of New York, in 1956, and this provided a welcome injection of capital at a time when blended Scotch was taking off worldwide. Ownership of the Scottish distilleries was transferred to Long John Distillers Ltd, and Glenugie was completely refurbished with new plant and equipment, two new stills and shell-and-tube condensers. Output was doubled. The company also built new distilleries at Tormore on Speyside and within Strathclyde, named Kinclaith.

In 1962 the company bought Laphroaig Distillery (see entry), and in 1970 changed its name to Long John International; in 1975, it was sold to Whitbread & Company Ltd, the brewers, whose spirits interests were bought by Allied-Lyons plc in January 1990 for £454 million. But by this time Glenugie had been closed for seven years. Not long after 1983 the property was divided and acquired by two North Sea oil engineering firms. The original buildings have since been demolished.

CURIOSITIES: The main distillery building had unusual cast iron framing. The distillery only ever had one pair of stills.

Glenury Royal Demolished

REGION
Highland (East)
ADDRESS
Glenury Road, Stone-
haven, Aberdeenshire
LAST OWNER
D.C.L./S.M.D.
CLOSED
1985

Joseph Hobbs landscaped the site and planted flowering shrubs and trees. He also installed a small laboratory at Glenury, with a pair of miniature stills for experimental distillations.

HISTORICAL NOTES: There was a distillery on the Ury estate before 1825, when the *Aberdeen Journal* reported that fire had destroyed its maltings, and in 1838 another Aberdeen newspaper states: 'The Glenury Distillery was originally established by the late Duke of Gordon, with a view to put down smuggling', but it does not say when.

It first appears in Excise records for 1833, owned by Captain Robert Barclay (1779–1854), Laird of Ury (his forbear had bought the estate in 1648) and sometime M.P. for Kincardine. Barclay was a Quaker and a progressive farmer, and built (or rebuilt) the distillery to provide a market for local barley. He also had a friend at Court, whom he referred to as 'Mrs Windsor', who persuaded King William IV to allow him to use the 'Royal' suffix from 1835.

On his death in 1847 the distillery was sold to William Ritchie of Dunottar and remained in his family until 1936, when it was sold to Joseph Hobbs for £7,500. It had been closed since 1925, but resumed the following year, and in 1938 Hobbs sold it on to Associated Scottish Distilleries Ltd for £18,500. This company was a subsidiary of National Distillers of America, which had made an arrangement with Hobbs to acquire distilleries: between 1934 and 1938 they bought Bruichladdich, Glenesk, Fettercairn, Glenlochy, Benromach and Strathdee Distilleries, as well as Glenury Royal, which became the company's head office. (See *Ben Nevis*.) National Distillers withdrew from Scotland in 1953 and sold A.S.D. and its distilleries to the D.C.L. (Bruichladdich and Fettercairn had already been sold, and Strathdee closed).

In 1965/66 Glenury Royal was expanded to four stills and much rebuilt.

It was terminally closed in 1985 and the site sold for residential development eight years later.

CURIOSITIES: Captain Barclay was an exceptionally strong man, and a considerable athlete. In 1799 he walked from London via Cambridge to Birmingham – a distance of 150 miles – in two days; two years later he walked from Ury to Boroughbridge in Yorkshire in five days. He was a popular man. Two hundred of his neighbours entertained him to dinner in the distillery's maltings in 1838.

GLENURY ROYAL

36
YEAR OLD

Single Highland Malt Scotch Whisky
Distilled in 1970

70cl

Harris

REGION
Highland (Island)
ADDRESS
Tarbert, Isle of Harris
PHONE
01859 502212
WEBSITE
www.harrisdistillery.com
OWNER
Isle of Harris Distillers Ltd
VISITORS
Visitor centre
CAPACITY
230,000 L.P.A.

HISTORICAL NOTES: Anderson Bakewell and his family have long connections with Harris and a deep love for the place and its people. Indeed, to quote the distillery's website, 'The concept for the distillery grew out of a realisation that the island's natural assets could best be harnessed to address its acute economic problems through a project that brings this special place and its qualities to the attention of a wider audience.'

With the advice of industry veteran Simon Erlanger, former Sales & Marketing Director at Glenmorangie and now Managing Director of Isle of Harris Distillers, and the support of a £1.9 million grant from the Scottish Government and a further £0.9 million from HIE (towards the estimated cost of £10 million), building commenced in spring 2014, and the 'The Social Distillery' went into production on 17 December 2015. It is highly artisanal, filling an average of only two casks a day, matured and bottled on site and closely connected to the local community.

As well as malt whisky, Isle of Harris makes an excellent gin using sugar kelp among its botanicals. Including visitor centre staff, the distillery will employ twenty people.

CURIOSITIES: The first legal distillery in the Isle of Harris. In keeping with the 'social distillery' ethos, the team of four distillers are all locals new to the craft and the nosing panel is made up of six community volunteers as well as the distillery team. Spent draff is donated to a consortium of local crofters.

RAW MATERIALS: Malt from independent maltsters, peated to around 15ppm phenols. Soft water from Abhainn Cnoc a'Charrain.

PLANT: Traditional semi-Lauter mash tun (1.2 tonnes). Five Oregon pine washbacks (6,000 litres). One plain wash still (7,000 litres charge), one plain spirit still (5,000 litres charge). Shell-and-tube condensers. Stills from Frilli Impianti, Siena, Italy.

MATURATION: Primarily barrels from Buffalo Trace Distillery, with some ex-Oloroso sherry butts. Warehouse located on the western (windward) coast of Harris, to take advantage of the extreme weather conditions there.

STYLE: Mid-peated, full-bodied spirit.

Highland Park

REGION
Highland (Island)
ADDRESS
Holm Road, Kirkwall,
Orkney
PHONE
01856 874619/885632
WEBSITE
www.highlandpark.co.uk
OWNER
Edrington Group Ltd
VISITORS
Visitor centre and shop
CAPACITY
2.5m L.P.A.

HISTORICAL NOTES: Highland Park Distillery stands on a hill overlooking Kirkwall, Orkney's main town – on an area of common land long known as the High Park of Rosebank. It proudly describes itself as 'The Northernmost Scotch Whisky Distillery in the World'.

From 1798, illicit whisky had been made here by a local brewer named Mansie (i.e. Magnus) Eunson, described as 'the greatest and most accomplished smuggler in Orkney', and also as a 'thug and small-time hood'. He enjoyed the protection of the Provost (i.e. mayor) of Kirkwall, whose son took over the running of the enterprise until 1814, when he was forced to close.

In 1813 Rosebank was enclosed, divided into lots and its name changed to Highland Park (although the distillery itself was known as Kirkwall Distillery until 1876). Ironically, the lots which contained the distillery were taken up by the local Collector of Excise, John Robertson, who had long pursued Mansie Eunson (and who finished up as Supervisor of Excise in London). Another lot was taken by Robert Borwick, his son-in-law, and in 1826 Borwick acquired the whole ground, including the distillery, and immediately took out a licence.

Borwick ran the distillery until his death in 1840, when it passed to his son George. By 1860 George was 'tired of the trade' and leased the plant to a local firm; on his death in 1869, ownership passed to his son, the Reverend James Borwick, who put it on the market for £450, believing that owning a distillery was incompatible with his calling. It was sold to a local farmer, who sold it on in 1876 to William Stuart, owner of Miltonduff Distillery.

Stuart went into partnership with his cousin, James Mackay, and they set about improving Highland Park. During their first season they produced 19,300 gallons of whisky; this had more than doubled by 1882/83. Robertson & Baxter, brokers in Glasgow, became their main customer.

Upon receiving samples in 1914, Sir Alexander Walker wrote: 'I am in the process of conversion to the idea that Highland Park is the only whisky worth drinking and Johnnie Walker only fit for selling to deluded Sassenachs.'

James Mackay died in 1885, and Stuart now went into partnership with James Grant, who became sole partner ten years later.

He immediately replaced the two stills with larger ones, and installed two further pairs in 1897. But the boom of the 1890s was about to turn to bust. Orders from R. & B. dropped from 60,000 to a mere 107 gallons in 1904/05. But the distillery remained in production throughout World War I.

James Grant had brought his son and son-in-law into the business in 1908. His son Walter Grant turned the firm into a limited company and offered it to Robertson & Baxter's sister company, Highland Distilleries. The deal was concluded in 1937.

Highland Distilleries began to promote Highland Park as a single malt in 1979 and installed a visitor centre in 1986, which was awarded five stars by VisitScotland in 2000. The owning company's name was changed to Edrington in 1999.

CURIOSITIES: The *Pembroke Castle* called at Kirkwall on her maiden voyage in September 1883. Sir Donald Currie, the ship's owner, and his guests were entertained by the town's leading councillor, '. . . who produced his well-known big bottle of Old Highland Park whisky. No sooner had this famous brand been tasted than they one and all agreed they had never met with any whisky like it before, that what was called Scotch in England was as different from this as chalk from cheese.' Twelve gallons of Highland Park were taken aboard, and the cruise continued to Copenhagen, where '. . . the King of Denmark, the Emperor of Russia and a very distinguished party were entertained on board. The Highland Park whisky was procured and pronounced by all the finest they had ever tasted.'

A hogshead of Highland Park distilled in 1877 set a record price at auction in Edinburgh in 1892.

Immediately after acquiring Highland Park, Highland Distilleries were obliged to buy 'a quarter acre of supposedly valueless and unworked blue stone quarry – Cattie Maggie's Quarry', in order to secure their water supply. The pool in the quarry feeds the springs that 'give Highland Park its unique flavour'.

Edrington spends around £20 million per annum on sherry casks for Highland Park. Since 2012 several new expressions of this malt have been released.

RAW MATERIALS: Hard water pumped up to the distillery from the Crantit Lagoons and one spring. Own floor maltings produce 20% of requirements, peated at around 20ppm phenols; the rest is unpeated from independent maltsters. Local peat from Hobbister Hill.

PLANT: Stainless steel full-Lauter tun with a peaked canopy (5.5 tonnes per mash), 12 Oregon pine washbacks and two of Siberian larch. Two plain wash stills (14,600 litres charge each), two plain spirit stills (9,000 litres charge each) both indirect fired by steam. Shell-and-tube condensers.

MATURATION: 25% first-fill ex-sherry, 60% refill ex-sherry, 15% refill hogsheads. Nineteen dunnage and four racked warehouses on site holding over 45,000 casks; a further 62–63,000 casks matured elsewhere.

STYLE: Malty, slightly smoky.

MATURE CHARACTER: The nose presents heather pollen and liquid honey, with caramelised oranges, sweet malt, a hint of oak and a drift of smouldering heather. The mouthfeel is smooth, the taste sweet, slightly salty then dry, with toffee, traces of spice (cinnamon, ginger) and a twist of smoke. Medium-bodied.

Imperial Demolished

REGION
Speyside
ADDRESS
Carron, Moray
LAST OWNER
Chivas Brothers
VISITORS
No

A local journalist reported of Imperial's crown that 'once gilded it would flash and glitter in the sunlight like the crescent on a Turkish minaret … among the dark pine woods of Carron and the brown hills which encircle the rushing Spey'.

HISTORICAL NOTES: Built in 1897 by Thomas Mackenzie, owner of nearby Dailuaine Distillery and distant Talisker on the Isle of Skye, Imperial's construction (an iron framework supporting thick red-brick walls, to resist fire) was a novelty in Scotland. The distillery was named in honour of Queen Victoria's Diamond Jubilee, celebrated that year, and the original maltings (designed by Charles Doig) was surmounted by an enormous cast iron crown.

This was not a good time to open a distillery, since the Whisky Boom of the 1890s burst in 1899. Imperial closed after one season. Production restarted in 1919, but difficulties with effluent disposal caused it to be closed again in 1925, by which time it was owned by D.C.L. The maltings continued to operate, but it was not for 30 years that improvements in waste treatment combined with demand to allow it to re-open, and in that year (1955) the mashhouse and stillhouse were modernised. A second pair of stills and Saladin maltings were added in 1964.

S.M.D. closed Imperial in 1985 and sold the distillery to Allied four years later. They refurbished and went back into production in 1991, but it was mothballed again seven years later. In truth, the distillery has been closed for 60% of its existence.

When Allied was broken up and sold in 2005, Imperial Distillery went to Chivas Brothers. Pernod Ricard announced in October 2012 that Imperial Distillery (closed St Andrew's Day 1998) would be demolished and rebuilt. It has been replaced by the strikingly modern Dalmunach Distillery (see entry).

CURIOSITIES: It is said that one of the problems with Imperial was the size of its four stills – so large they cannot be run flexibly. The problem of effluent disposal was solved in the 1950s when it was discovered that nutritious elements in draff and pot ale could be recovered by drying, to make a high-protein animal feed.

The new distillery was opened in June 2015.

InchDairnie

In

REGION
Lowland

ADDRESS
Whitecraigs Road,
Glenrothes, Fife

PHONE
+44(0)1592210014

WEBSITE
www.inchdairnie
distillery.com

OWNER
John Fergus & Co.

VISITORS
No

CAPACITY
2 million L.P.A.

HISTORICAL NOTES: The seven hectare site chosen for this substantial new distillery once belonged to the InchDairnie Estate, near Kinglassie in Fife. The name is believed to mean 'secret land close to the water' – a combination of 'Inch', a stretch of low-lying land close to a river, loch or sea and 'Darne', 'secret, concealed or hidden place'.

With a £1.6 million grant from the Scottish Government, £240,000 from Scottish Enterprise, private equity from Denmark and a strategic partnership with Glasgow blenders MacDuff International, construction began in July 2014, with production starting in December 2015.

The distillery is strikingly contemporary in appearance, housed in a large, angular, slate-grey building, lit by a 'wall of glass' at one end. Adjacent is an office/reception building, part clad with grey stone and timber panels. There are two large grey warehouses on site, with room for a further six,

The project is the brain-child of Ian Palmer, Managing Director and founder (2011) of the owning company, John Fergus & Co. Ian started his career as a process engineer at Invergordon Distillers and has since worked as Distilling General Manager at Whyte & Mackay, Operations Director at JBB and Kyndal International and as the General Manger at the Glen Turner Company Ltd., where he was responsible for the construction of Starlaw Distillery (see entry).

He has designed a highly innovative production system for Inchdairnie, with a view to exploring and enhancing flavour, reducing waste and saving energy.

The malt is pulverised to flour in a hammer mill and processed in a Belgian-made mash conversion vessel with 22 plates, which increases the extraction of very clear worts. Only Teaninich Distillery (see p. 348) employs a similar system. The stills employ a complex 'thermal vapour recompression system' with two condensers on each still which both increases the copper contact and, by allows for a 40%–50% saving on energy costs by recycling the heat to drive the stills.

Using different yeasts, two core spirits are being made: 'InchDairnie' will be bottled as the distillery's own single malt and 'Strathenry' will be traded with other distillers, especially the distillery's strategic partner, MacDuff International, owner of the Lauder's, Islay Mist and Grand McNish blends. MacDuff will store some of its stock on site and have a blending facility there. In the first year, InchDairnie distillery is aiming to produce two million litres, but has the capacity to double this. This will be stored in 44,000 casks in two warehouses, with an additional seven warehouses planned for the future.

RAW MATERIALS: Locally grown winter and spring barley, malted locally. The use of winter barley is unusual, and the idea of creating distinct spirits from each is unique. Water from the Flowers of May and the Goatmilk Hill springs.

PLANT: A hammer mill pulverises the malt, which is then processed in a Meura mash filter (see Teaninich) using a mix of yeast strains and fermented in 4 external stainless steel washbacks at higher than normal gravities. One six-plate Lomond still (at present not connected, but may be used for triple distillation or as an alternative spirit still), one Frilli (Italian) wash still 18,000 litres charge), one Frilli spirit still (11,000 charge)

MATURATION: A mix of first fill and re-fill ex-Bourbon, ex-sherry butts and ex-wine barrels.
All matured on-site, currently in two warehouses each holding 44,000 casks. Planning permission for a further seven warehouses has been granted.

STYLE: Full-bodied and complex spirit with a slightly sweet edge.

Inchgower

REGION
Speyside

ADDRESS
Buckie, Moray

PHONE
01542 836700

WEBSITE
www.malts.com

OWNER
Diageo plc

VISITORS
By appointment

CAPACITY
3.2m L.P.A.

HISTORICAL NOTES: Inchgower stands outside the fishing port of Buckie. The distillery was built in 1871 by Alexander Wilson & Company to replace their distillery at Tochineal nearby, which had become too small – and where the landlord had doubled the rent. Tochineal's plant was moved over to the new distillery. Wilson & Company operated Inchgower until 1936, when the firm went bankrupt and both the distillery and the family house were bought by Buckie Town Council for £1,600; two years later the distillery was sold for £3,000 to Arthur Bell & Sons.

Bell's was acquired by Guinness in 1985, Guinness bought the D.C.L. in 1987, and thus Inchgower came to be owned by Diageo.

Inchgower was closed for 16 weeks in 2012 for upgrading. Combined with operating 51 weeks a year, this has increased its capacity from 1.9 million to 3.2 million L.P.A. per annum.

CURIOSITIES: Inchgower was Bell's third distillery (the first two were Blair Athol and Dufftown). It was bought by Arthur Kinmond Bell, Chairman of the company and oldest son of Arthur Bell. A.K. was a notable philanthropist, and during the 1930s had built 150 high-quality houses and offered them at low rent to needy and unemployed people in Perth. He died four years after acquiring Inchgower, in 1942. The *Perthshire Advertiser* referred to him as Perth's 'greatest benefactor of all time'.

The distillery supplied fillings for Bell's blends, and as the brand's popularity increased it was expanded (to four stills in 1966) and worked harder and faster, which did not help the quality of the make.

From 1979 Bell's was the No. 1 best-selling Scotch in the U.K. (a position it holds to this day).

Like all distilleries, Inchgower operated a farm as part of its business. In 1885 the residues from the distillery (draff and pot ale) fed 100 head of cattle, 200 sheep and pigs.

RAW MATERIALS: Unpeated malt from Burghead. Water from springs in the Minduff Hills.

PLANT: Mash tun (8.2 tonnes). Six Oregon pine washbacks. Two plain wash stills and two plain spirit stills. All indirect fired. Shell-and-tube condensers, with after-coolers.

MATURATION: Mainly refill American oak.

STYLE: Nutty-spicy, malty.

MATURE CHARACTER: The nose is malty, caramelised and lightly sherried, yet the overall impression is dry. There are some coffee and chocolate notes, and sometimes a whiff of smoke. The taste is sweet then dry, with a hint of malt, tinned apples and hazelnuts. Medium-bodied.

In

Invergordon Grain Distillery

ADDRESS
Invergordon, Ross-shire
PHONE
01349 852451
OWNER
Whyte & Mackay Ltd
CAPACITY
32m L.P.A.

HISTORICAL NOTES: By 1958 all government restrictions on whisky distilling had been relaxed. There was an acute shortage of capacity, in both malt and grain spirit, to meet the burgeoning post-war demand. The grain whisky distilleries at Lochside, Girvan, Moffat and Invergordon are examples of the industry's response to this demand.

The port of Invergordon stands on the north shore of the Cromarty Firth, 'a place of considerable mark', according to the *Imperial Gazetteer* (1854), 'substantially built, well-situated for traffic and of growing importance for the shipment of farm produce from the surrounding country'. At that time it had a population of nearly 1,000. The port was named for Sir William Gordon, its owner in the late eighteenth century.

Since World War II, several attempts had been made to bring industrial activity to the Highlands north of Inverness. One such was Invergordon Distillery, promoted strongly by James Grigor, Provost of Inverness, in the late 1950s. With good reason: communications by sea and road were excellent; it was on the edge of a notable barley-growing region, and the water was first-rate.

The Invergordon Distillers Ltd was incorporated in March 1959 to build the first and only grain whisky distillery in the Highlands. Production commenced in July 1961, with one Coffey still producing 25,950 litres of pure alcohol per week; Stanley P. Morrison was appointed agent. Two further Coffey stills were added in 1963, and another in 1978, with extra columns on the Dumbarton model (see entry) to produce neutral spirit. These were designed in-house by the distillery's engineer.

In 1985 Invergordon bought the long established whisky firm Charles Mackinlay & Company (see *Jura, Glenallachie* etc.) for £7.8 million, and three years later four of the directors led a management buy-out. Their independence did not last long: Whyte & Mackay (which had a 41% shareholding

In 1916, under the Defence of the Realm Act, Lloyd George established the state management of all pubs and off-licences in Invergordon and Carlisle as an experiment. This remained in Carlisle until the 1970s.

in the company), having failed to take over in 1991, acquired a sufficient majority in October 1993.

The subsequent history of Invergordon Distillery is linked to the ups and downs of Whyte & Mackay (see *Jura* and 'Who Owns Whom?' in *Facts and Figures*) In 2007, Whyte & Mackay, the owner of Invergordon, was bought by United Spirits Ltd (U.S.L.), a division of the United Brewers Group, the largest brewer and distiller in India, for £595 million.

In 2013, Diageo bought a 28% stake in U.S.L., and later announced its intention to acquire a further 26%. The U.K. Office of Fair Trading said Diageo's control of U.S.L. was against competition law and could lead to higher whisky prices in the U.K. In response, Diageo said it would sell Whyte & Mackay.

In May 2014, the Philippines-based brandy distiller Emperador bought Whyte & Mackay and its distilleries for £430 million. Emperador is the second best-selling spirit in the world, 31,950,300 cases sold in 2013 (Johnnie Walker, the world's best-selling Scotch whisky, sold 19,288,300 cases).

CURIOSITIES: In 1965 a pot-still malt distillery named Ben Wyvis was built within the Invergordon complex. It ceased production in 1977 (see *Ben Wyvis*).

From 1990 Whyte & Mackay released an Invergordon Single Grain, but this has now been discontinued.

RAW MATERIALS: Water from Loch Glass.

PLANT: At the outset, Invergordon had a single column still. Two more stills were added in 1963 and another (much larger) one in 1978. This new, larger Coffey still was used for the production of neutral spirit.

MATURATION: American oak. On site.

STYLE: Light and sweet.

MATURE CHARACTER: Fresh and sweet, with pear drops and acetone. Clean and light-bodied, but somewhat one-dimensional compared to malt whisky.

Inverleven Demolished

REGION
Lowland

ADDRESS
Glasgow Road,
Dumbarton

LAST OWNER
Allied Distillers

CLOSED
1991

HISTORICAL NOTES: Inverleven was installed within Hiram Walker's Dumbarton grain whisky distillery in 1938, the same year that the distillery opened. Both were licensed to Walker's subsidiary George Ballantine & Sons, designed to make fillings for the Ballantine's blends.

Inverleven had two conventional stills and (after 1959) a Lomond still; the latter has disappeared, but the former two were bought by Bruichladdich Distillery in 2007 and will be used in their proposed Port Charlotte Distillery on Islay.

The stills were originally direct fired, but were changed to steam heating in the early 1960s. Inverleven closed in 1991. It had a capacity of 1.3m L.P.A.

CURIOSITIES: The distillery's site, in the shadow of Dumbarton Rock and at the mouth of the River Leven, was formerly a shipyard (derelict since 1933), and the high red-brick building itself was designed in America. It is 'very much what one would expect to find in the Midwest of the United States rather than Lowland Scotland' (Philip Morrice). The building was demolished in 2006.

'It was said that the plant might not have been built had not Sir Henry Ross, Chairman of D.C.L., kept Harry Hatch of Hiram Walker waiting overlong, whereupon Hatch, who had come to buy grain whisky, announced that he would build his own grain distillery.' (Charles Craig)

Inverleven was never bottled by its owner and only occasionally bottled by independents.

Jura

REGION
Highland (Island)
ADDRESS
Craighouse, Isle of Jura,
Argyll
PHONE
01496 820240
WEBSITE
www.jurawhisky.com
OWNER
Whyte & Mackay Ltd
VISITORS
Visitor centre, and
apartments
CAPACITY
2.3M L.P.A.

HISTORICAL NOTES: Perhaps surprisingly for a place so remote from the Excise, Jura had a licensed distillery by 1810 – at Craighouse, the site of the present distillery and also of a previous, unlicensed still – named the Small Isles Distillery. The licensee was Archibald Campbell, the laird of Jura.

Several tenants ran the distillery with little success. The first was William Abercrombie (to 1831), then Archibald Fletcher (until 1851). When Fletcher gave up the lease, there were a mere 5,450 litres of whisky under bond. Campbell's son thought of selling the stills for scrap (value £400), but was soon approached by Norman Buchanan of Glasgow (who acquired Caol Ila Distillery at the same time), but he went bankrupt ten years later. After a gap of some years, in 1876 the laird signed a 34-year lease of the distillery with James Ferguson of Glasgow, but he abandoned the site in 1901 (taking with him all the equipment, which he had paid for). The laird removed the roof to avoid paying rates and the original distillery fell into ruin.

It was revived in 1963 by two local landowners, Robin Fletcher and Tony Riley-Smith (uncle of the founder of *Whisky Magazine*), who wanted to bring employment to the island. With backing from Mackinlay Macpherson & Company (soon taken over by Scottish & Newcastle Brewers), they recruited the services of William Delmé-Evans, who had designed and built Tullibardine Distillery in the late 1940s. 'It was our intention to produce a Highland-type malt, differing from the typically peaty stuff last produced in 1900,' he wrote. The first single malt bottling was released in 1974.

The distillery was expanded by Dr Alan Rutherford (who went on to become Head of Production at U.D.) from 1976 to 1978 (from two to four stills). Invergordon Distillers acquired Mackinlay's and its distilleries in 1985, and were themselves bought by Whyte & Mackay ten years later.

George Orwell lived in a remote cottage on Jura in the late 1940s, and finished Nineteen Eighty-Four *here.*

CURIOSITIES: The stills here, designed by Delmé-Evans, are unusually tall (7.7m), to produce a lighter style of malt.

Apartments in the former manager's house were refurbished in 2006 and are available to let. Jura has long been packaged in an unusual waisted bottle, elegant and even voluptuous. This bottle shape was first adopted by MacKinlay & Birnie for their Glen Mhor malt, and has been used by Jura from the start.

Sales of Jura single malt have increased dramatically in recent years, from 350,000 bottles per annum in 2010 to 1.9 million bottles in 2014 – the third best-selling malt in the UK. This is partly owing to the skill of Whyte & Mackay's legendary Master Blender, Richard Paterson, who selects the casks.

RAW MATERIALS: Unpeated malt and some peated malt, and also malt from Port Ellen Maltings, Inverness, Aberdeen and Pencaitland. Water from the Market Loch.

PLANT: Semi-Lauter mash tun (4.75 tonnes); six stainless steel washbacks. Two wash stills (25,000 litres capacity each, 24,150 litres charge) of lamp-glass design; two lamp-glass spirit stills (22,000 litres capacity each, 15,500 litres charge). Indirect fired. Shell-and-tube condensers.

MATURATION: 27,000 casks matured on site in high racked warehouses (for single malt bottlings). 50% first-fill bourbon, 50% second- and third-fill; 5% sherry butts used in some expressions. 33% of the make bottled is as single malt.

STYLE: Oily, earthy-piney, with lemon notes.

MATURE CHARACTER: The oiliness apparent in the new-make comes through in the mature whisky, along with pine sap, orange zest and dry nuts. The taste is oily, with some maltiness; sweetish to start, becoming dry, with a dash of salt. Medium- to light-bodied.

Ki

Kilchoman

REGION
Islay

ADDRESS
Rockside Farm, Isle of
Islay, Argyll

PHONE
01496 850011

WEBSITE
www.kilchomandistillery.
com

OWNER
Kilchoman Distillery
Company Ltd

VISITORS
Visitor centre

CAPACITY
200,000 L.P.A.

HISTORICAL NOTES: Kilchoman's situation on the wild west coast of Islay, tucked in behind the sea cliffs and close to the wide white strand of Machir Bay, is striking. It is also historic. These lands were gifted by the Lords of the Isles to their physicians, the Beaton or MacBeatha family, who arrived in Scotland from Ireland in 1300, and may well have brought the secrets of distilling with them. Fergus McVey was confirmed in possession by James VI in 1609: the fine carved cross in Kilchoman kirkyard, dating from even earlier, commemorates another member of the family.

The distillery was the brain-child of Anthony Wills, wine and spirits merchant, who moved to Islay in 2000, having married Kathy Wills, an Illeach, some years earlier. A year later he leased and renovated some semi-derelict buildings at Rockside Farm, Kilchoman, from its owner, Mark French, who grew malting-quality barley and had a fine herd of cows to consume the residues of distilling. The distillery was officially opened by the present writer on 3 June 2005, and went into production in December that year.

CURIOSITIES: Kilchoman's slogan is 'Taking whisky back to its roots', based on the fact that it is a veritable farm distillery, as many were in days gone by – on Islay alone there were 13 farm distilleries in the early nineteenth century. It is not unique in this regard (see *Daftmill*), however it can make the proud claim of doing everything on site – from growing the barley to bottling the whisky – and this is unique.

The distillery has a charming café and shop run by Kathy Wills. The whole enterprise has a family feel about it.

The highly experienced and well-known distillery manager, John McLellan, moved to Kilchoman from Bunnahabhain in 2010, and that year Kilchoman was named 'Distillery of the Year' by *Malt Advocate*,

It might be argued that Kilchoman was the cradle of distilling in Scotland.

America's leading whisky magazine. In 2015, Kilchoman was able to secure its future and its supply of malting barley by buying Rockside Farm.

RAW MATERIALS: Soft peaty water from the Allt Glean Osmail Burn. Floor maltings on site, using barley grown on the farm and peat from Duich Moss (30ppm phenols) and the balance from Port Ellen Maltings (heavily peated at 50ppm phenols). Own grown malt accounts for 30–40% of production – spirit made from this is casked separately from that made from Port Ellen malt.

PLANT: Semi-Lauter mash tun (one tonne). Four stainless steel washbacks. One plain wash still (2,700 litres charge), one boil-pot spirit still (1,500 litres charge), both indirect fired by steam coils. Shell-and-tube condensers.

MATURATION: 85% in fresh (currently from Buffalo Trace) and refill ex-bourbon barrels, and 15% in ex-Oloroso butts. All matured on site.

STYLE: Unusually sweet, fruity and smoky.

MATURE CHARACTER: The nose is maritime, often with light dried fruit notes; the taste sweet to start, with some saltiness and acidity and a smoky finish.

Kinclaith Demolished

REGION
Lowland
ADDRESS
40 Moffat Street,
Glasgow
LAST OWNER
Allied Distillers
CLOSED
1975

HISTORICAL NOTES: Kinclaith was built in 1956/57 within Strathclyde Distillery to provide fillings for Long John (see *Strathclyde*). It was dismantled in 1976/77, after Long John International had been sold to Whitbread, to allow for expansion at Strathclyde.

A miniature bottle (5cl) of Kinclaith 1966 achieved £107 at auction in 2002.

Never bottled by its owner, a handful of bottlings have been released by G. & M., Cadenhead's and Signatory – all of which achieve high prices at auction.

Kingsbarns

REGION
Lowland

ADDRESS
East Newhall Farm,
Kingsbarns, St Andrews,
Fife

PHONE
07717 754053

WEBSITE
www.kingsbarnsdistillery.
co.uk

OWNER
Wemyss family

VISITORS
Visitor centre

CAPACITY
600,000 L.P.A.

HISTORICAL NOTES: The idea of converting a picturesque farm steading on Sir Peter Erskine's Cambo Estate near Kingsbarns was conceived by local golf caddie, Doug Clement, in 2009. After building a professional team including Edinburgh architects Simpson and Brown, as well as Bill Lark, a Tasmanian distiller, Clement then raised some initial seed capital from a number of small investors and secured planning permission from Fife Council.

In late 2012, Clement received a £670,000 grant from the Scottish Government which in part enabled him to bring on board the Wemyss family, who took over the project in early 2013 with only Clement remaining on board as founding director and visitor centre manager. The new distillery manager is Peter Holroyd, previously head brewer at Strathaven Brewery. The Wemyss family's ancestral home is also in Fife, at Wemyss Castle, and the family have been independent bottlers of Scotch whiskies since 2005 through their Wemyss Malts brands based in Edinburgh.

The farm steading, comprising a doocot, horse-mill and variety of old barns, dates from the late eighteenth century, and now houses a comfortable visitor centre as well as the distillery itself. Kingsbarns went into production on St Andrew's Day (30 November) 2014, and on its first anniversary a Founders' Club was established (see their website).

RAW MATERIALS: Local barley from the East Neuk of Fife, malted without peating by independent maltsters. Soft water from a 100m bore-hole directly beneath the distillery.

PLANT: Stainless steel semi-Lauter mash tun (1.5 tonnes). Four steel washbacks, with room for a further two. One wash still (7,500 litres charge) and one spirit still (4,500 litres charge), indirect fired. Shell-and-tube condensers.

MATURATION: Mainly refill ex-bourbon casks, off site.

STYLE: Lowland; fresh and grassy.

Ki

Kininvie

REGION
Speyside

ADDRESS
Dufftown, Moray

PHONE
01340 820373

WEBSITE
No

OWNER
William Grant & Sons Ltd

VISITORS
No

CAPACITY
4.8m L.P.A.

A limited bottling (500 bottles) was released in February 2008 to mark the opening of Heathrow Terminal 5.

HISTORICAL NOTES: Kininvie Distillery was built from scratch in 1990 next door to William Grant's Glenfiddich/Balvenie Distilleries. It went into production on 4 July. The distillery was mothballed in December 2009 for stock adjustments, possibly connected to the opening of Ailsa Bay the previous December (see entry), however it resumed production in 2012, with capacity expanded by increased shifts to 4.8 million L.P.A. (from 4.4 million).

CURIOSITIES: It was opened by Janet Sheed Roberts, William Grant's grand-daughter and the oldest woman in Scotland at the time of her death in April 2012, when she was 111. Grant's released 11 bottles of Glenfiddich Janet Sheed Roberts to mark her 110th birthday: the first was sold by Bonhams in December 2011 for £46,850 and another in New York in March 2012 for £59,252. All the proceeds went to charities.

Limited amounts of Kininvie have been released at 17YO as Hazelwood Reserve (the name of Janet Sheed Roberts' home close to the distillery).

In order to prevent independent bottlers using their brand-names, Kininvie, Balvenie and Glenfiddich, William Grant 'tea-spoon' (i.e. add small amounts of one make into another) before selling casks, and give them different names. Aldunie (Kininvie), Burnside (Balvenie) and Wardhead (Glenfiddich).

RAW MATERIALS: Water from springs in the Conval Hills. Unpeated malt from independent maltsters.

PLANT: Semi-Lauter mash tun located within Balvenie Distillery, but dedicated to Kininvie's needs (11.25 tonnes). Nine Douglas fir washbacks. Three plain wash stills (14,600 litres charge), six boil-pot spirit stills (8,600 litres charge) all indirect fired by steam coils. Shell-and-tube condensers.

MATURATION: Refill ex-bourbon and ex-sherry casks.

STYLE: Floral.

Knockando

REGION
Speyside

ADDRESS
Knockando, Aberlour,
Moray

PHONE
01479 874660

WEBSITE
www.malts.com

OWNER
Diageo plc

VISITORS
By appointment only

CAPACITY
1.4m L.P.A.

HISTORICAL NOTES: The distillery was designed by Charles Doig for John Tyler Thomson, chartered accountant and spirits broker in Elgin, who traded as the Knockando-Glenlivet Distillery Company. It is situated on the north bank of the Spey, and the Strathspey railway runs through the distillery; a dedicated siding was installed in 1905. The name is that of the parish, and comes from Cnoc-an-Dubh, 'the dark hillock'.

Production started in 1899, the year of the collapse of Pattison's of Leith, a leading blender and buyer of fillings. With the whisky industry in recession as a result, Knockando closed the following year, and was sold in 1904 to W. & A. Gilbey, the London wine and spirits merchant, for £3,500.

Gilbey's merged with United Wine Traders (part of which was Justerini & Brooks) in 1962, to become Independent Distillers & Vintners, and Knockando became the key constituent of J. & B. Rare, and later the 'brand home' (which it remains). In 1969 it was doubled in size (to four stills).

I.D.V. was taken over by Watney Mann in 1972, who sold to Grand Metropolitan the same year. Grand Metropolitan/I.D.V. and Guinness/U.D. merged in 1997 to form U.D.V., which in turn became Diageo.

CURIOSITIES: I.D.V. were offering Knockando as a single malt by 1977/78. As wine merchants, they also bottled by 'vintage' rather than age, although the 'vintages' declared were usually around 12 years. This practice is now discontinued.

From 1978 until 2006, Knockando was managed by Innes Shaw, who had worked there man and boy. His great-grandfather was employed as a joiner when the distillery was built, and the visitor centre exhibits invoices from him. He later became Manager at Cragganmore and an exciseman.

Knockando was the first distillery on Speyside to install electricity.

RAW MATERIALS: Soft process water from the Cardnach Spring; cooling water from the River Spey. Floor maltings until 1968; now lightly peated malt from Burghead.

PLANT: Full-Lauter mash tun (9.5 tonnes). Four Oregon pine washbacks. Two lamp-glass wash stills (16,000 litres charge); two boil-ball spirit stills (16,000 litres charge). All indirect fired. Shell-and-tube condensers.

MATURATION: Refill ex-bourbon hogsheads.

STYLE: Malty, cereal-like.

MATURE CHARACTER: The keynote breakfast-cereal style comes through in the standard 12-year-old. Sugar Puffs to be exact, with honey, walnuts and a trace of olive oil. The taste is sweet and simple, with cereal and nuts. Medium- to light bodied. A good breakfast malt!

Knockdhu

REGION
Highland (East)

ADDRESS
Knock, by Keith, Moray

PHONE
01466 771223

WEBSITE
www.ancnoc.com

OWNER
Inverhouse Distillers Ltd

VISITORS
By appointment

CAPACITY
1.9m L.P.A.

HISTORICAL NOTES: The distillery is situated beneath the dark, rounded hump of Knock Hill: *Knockdhu* means 'the dark hillock'. It was the first malt whisky distillery to be commissioned by D.C.L., which until then, 1893/94, had confined its interests to grain whisky production. The site was chosen on account of the water quality from springs on the hill, close to 'good barley country' and near 'an inexhaustible supply of excellent peats'; it also helped that the Great North of Scotland Railway line between Aberdeen and Elgin ran adjacent to the site. The make was used primarily in the Haig blends.

A visitor remarked in 1925 how 'remarkably well kept' the site was, with a 'handsome avenue' leading up to the 'severe buildings', and a nicely laid-out orchard of apple, cherry and plum trees. Even at that time it was equipped with 'a powerful electric light plant'. In 1930 management was transferred to S.M.D.

Knockdhu was a casualty of the world recession of the early 1980s. It was closed in 1983, but was sold to Inver House five years later. They released the first single malt bottling in 1990, then changed the brand name to an Cnoc (see below). Inver House became part of the leading Far Eastern spirits company ThaiBev plc in 2001.

CURIOSITIES: When it was built, Knockdhu was a very modern distillery. Its refrigeration plant was the first of its kind in the north of Scotland; the Manager and Excise Officer were provided with 'handsome villas' and 'private lavatories'. Situated three miles east of Keith, it is on the very edge of Speyside, and some classify the make as such.

In 1960 a tractor replaced the horse and cart that moved goods between the distillery and the railway station. 'A sad day for everybody,' wrote an employee. 'And particularly for the man who worked with the horse, because it was a dear friend.'

The distillery is Knockdhu, the malt is an Cnoc,

Between 1940 and 1945 the distillery was occupied by a unit of the Indian army, billeted in the malt barns, with stabling for their horses and mules.

following a gentleman's agreement not long after Inver House took over, in order to avoid any possible confusion with Knockando. Its positioning is young-ish and artistic, with sponsorship focused on the arts.

RAW MATERIALS: Process water from springs on Knock Hill; cooling water from the Ternemny Burn. Malt from independent maltsters.

PLANT: Traditional deep bed, cast iron mash tun but with semi-Lauter stirring gear (4.15 tonnes). Six Douglas fir washbacks. One tall boil-ball wash still (10,500 litres charge), one tall boil-ball spirit still (11,000 litres charge). Both indirect fired. Worm tub shared by both stills.

MATURATION: Mainly U.S. ex-bourbon casks. Matured on site in four dunnage warehouses and one racked warehouse for single malt bottlings, holding some 7,600 casks.

STYLE: Speyside style – fruity-floral, estery, with lemon notes, but with added body from the worm tubs.

MATURE CHARACTER: A light Speyside style; sweet, floral (buttercups), cereal, lemony, with a suggestion of vanilla cream. The texture is light, the taste sweet, with cooked apples and lemon meringue pie.

Ladyburn Dismantled

REGION
Lowland
ADDRESS
Girvan, Ayrshire
LAST OWNER
William Grant & Sons Ltd
CLOSED
1975

HISTORICAL NOTES: William Grant & Sons, owners of the phenomenally successful Grant's Standfast blended whisky (now Grant's Family Reserve), built a large grain distillery at Girvan in 1966, with a small malt whisky distillery within it, installed the same year, to supply fillings for their blend (see *Girvan*).

The distillery had two pairs of stills. The venture was abandoned in 1975 and the distillery dismantled.

Ladyburn was the first fully automated malt whisky distillery in Scotland.

CURIOSITIES: According to Misako Udo (author of *The Scottish Whisky Distilleries*), Grant's still possess 30 casks of Ladyburn and may be bottling it as a single in the near future.

Lagavulin

REGION
Islay

ADDRESS
Port Ellen, Isle of Islay, Argyll

PHONE
01496 302749

WEBSITE
www.malts.com

OWNER
Diageo plc

VISITORS
Visitor centre

CAPACITY
2,45ml PA

HISTORICAL NOTES: Guarding the opening to Lagavulin Bay stand the crumbling remains of Dunyveg Castle, power-base of the Lords of the Isles and where the lords kept their galleys of war.

There were illicit stills in this sheltered 'hollow of the mill' in the eighteenth century, but the first licensed operation was established in 1816 by John Johnston. In the 1820s he acquired a neighbouring distillery, Ardmore, and his son combined the two in 1837.

John Crawford Graham, brother of the Glasgow wine merchant Alexander Graham and in partnership with James Logan Mackie, bought the distillery in 1852. Another brother, Walter Graham, had the lease of Laphroaig Distillery, next door, and ran the two together for four years. J.L. Mackie took over the business in 1860 and sent his nephew Peter to serve an apprenticeship at Lagavulin.

Peter Mackie joined his uncle in the business and in 1890 became Managing Partner. The same year, he created the blend White Horse; so successful was this that by 1908 Mackie & Company was named among the Big Five (with Walker's, Dewar's, Buchanan and Haig). In 1924, the year of Peter Mackie's death, the company became White Horse Distillers. This merged with D.C.L. in 1927.

CURIOSITIES: When Alfred Barnard visited Lagavulin in 1886 he described the 8YO whisky as 'exceptionally fine'. Later he wrote: "There are only a few of the Scotch distillers that can turn out spirit for use as single whiskies and that made at Lagavulin can claim to be one of the most prominent". To mark the distillery's bi-centenary in 2016, Diageo released an 8YO 'available for one year only'. Peter Mackie was an authority on shooting, and wrote *The Gamekeeper's Handbook*. He insisted his distillery workers ate a proper diet (see *Craigellachie*) and was a tireless spokesman for the whisky industry. He was made a baronet in 1920.

'There are only a few of the Scotch distillers that can turn out spirit for use as single whiskies, and that made at Lagavulin can claim to be one of the most prominent.'

Alfred Barnard, 1887

In 1908, Mackie built a second distillery within Lagavulin, which he named Malt Mill. Until the previous year he had been the agent for Laphroaig, but he fell out with the owners in considerable acrimony, and resolved to make a whisky which was very similar and which would damage Laphroaig's market. The plan was a failure, but Malt Mill operated until 1962; its make went into blends.

A (fictional) cask of Malt Mill was the centrepiece of Ken Loach's award-winning movie *The Angels' Share* (2012), and achieved over £1 million!

The building which housed Malt Mill is now Lagavulin's visitor centre.

RAW MATERIALS: Floor maltings until 1974; now heavily peated malt (30–35ppm phenols) from Port Ellen. Dark, soft water from the Solum lochs.

PLANT: Full-Lauter mash tun (4.32 tonnes). Ten Oregon pine washbacks. Two plain wash stills (10,500 litres charge), two plain spirit stills (12,200 litres charge). All indirect fired. Shell-and-tube condensers.

MATURATION: Mainly refill American, with some refill European. Some matured on site and at Port Ellen and Caol Ila Distilleries; mostly in the Central Belt.

STYLE: Rich, sweet, peaty.

MATURE CHARACTER: Lagavulin is the richest and most complex of the Islay malts. The nose has berries, Lapsang Souchong tea, sherry, sweet seaweed, wax polish, camphor, carbolic and scented smoke. A big, rich mouthfeel; very sweet to start, then a big blast of peat smoke in the finish and a long aftertaste. Full-bodied.

Laphroaig

REGION
Islay

ADDRESS
Port Ellen, Isle of Islay,
Argyll

PHONE
01496 302418

WEBSITE
www.laphroaig.com

OWNER
Beam Suntory

VISITORS
Yes, and a well-organised
Friends of Laphroaig
group

CAPACITY
3.3m L.P.A.

*John Campbell, the
distillery manager,
was born and raised
on Islay.*

HISTORICAL NOTES: Alexander and Donald Johnston, who were tenant farmers in Tallant and Kildalton by 1810, established the distillery at Laphroaig in 1815. Their forebears were MacIains from Glencoe (*Mac-Iain* Anglicises to 'John-son'), who came to Islay in the late fifteenth century.

Their landlord was Walter Frederick Campbell, an enlightened improver; he founded both Port Ellen, named after his wife, and Port Charlotte, named for his mother, and supported distilling ventures enthusiastically.

Donald Johnston bought his brother out in 1836, but died in 1847 after falling into a vat of pot ale and was succeeded by his 11-year-old son, Dugald. Laphroaig was looked after by his uncle and managed by Walter Graham from neighbouring Lagavulin Distillery until he came of age in 1857. Laphroaig labels still recall 'D. Johnston & Company', and the distillery remained in the family until the 1960s.

Perhaps the most remarkable scion of the family was Ian Hunter, great-grandson of Donald, who became Manager of Laphroaig in 1908 and sole owner of the business 20 years later. One of his first tasks was to sack the distillery's agents, Mackie & Company of Glasgow. Unusually, Laphroaig was being sold as a single malt even in those days. Peter Mackie (later Sir Peter), owner of Lagavulin Distillery, was furious; Mackie's had, after all, 'built the brand'. (See *Lagavulin*.)

In the 1920s, Ian Hunter set about selling his whisky in the United States, although at that time the sale of liquor was prohibited. While he was abroad, the distillery was managed by his secretary, Bessie Williamson, and when he died in 1954 he bequeathed Laphroaig to her.

By the mid 1950s the distillery was badly in need of repair. In order to raise the funds to do this, Bessie Williamson sold a third of her shares to an American distiller, the Schenley Corporation, and by 1970 Schenley had complete ownership. The days of privately owned distilleries were over and, like many others, Laphroaig became an item on a multi-national corporation's balance sheet: Long John International, Whitbread, Allied Lyons, Allied Domecq and, in 2005, Fortune Brands, owners of Jim Beam Bourbon.

Beam Inc. split from Fortune Brands in October 2011, and was bought by the Japanese distillery Suntory, in March 2014. The name changed to Beam Suntory (See *Ardmore*).

CURIOSITIES: Bessie Williamson was the first woman in modern times to manage a distillery (earlier there had been others, notably Elizabeth Cumming of Cardow).

Whitbread sold to Allied Lyons (in 1989), but not before appointing a manager at Laphroaig who would become a legend in the whisky trade – Iain Henderson. The new owners wanted to promote the brand vigorously, and Iain travelled the world as Brand Ambassador, increasing sales from 20,000 cases in 1989 to 170,000 cases in 2002, when he retired. He also introduced the 'Friends of Laphroaig' concept by which enthusiasts for the malt could acquire a one-foot-square plot of Islay (in a field behind the distillery). Currently 470,000 plots have been allocated, and friends come from 165 countries. They include The Prince of Wales, who granted his Royal Warrant to Laphroaig in 1994.

Laphroaig is the best-selling Islay malt, with sales increasing by around 20% per annum, to 277,425 nine-litre cases in 2014, the eighth best-selling malt in the world.

RAW MATERIALS: Soft peaty water from Kilbride Reservoir, with support from Loch na Beinne Brice. Own floor maltings supply around 15% of requirement; the rest from Port Ellen and mainland maltings (at 35–40ppm phenols).

Laphroaig was possibly the first single malt Scotch to be promoted in the U.S. Prohibition was still in place, but a loophole in the law allowed whisky to be sold 'for medicinal purposes'

Peat from Machrie Moss, owned by the distillery and still cut by hand.

PLANT: Stainless steel full-Lauter tun (8.5 tonnes per mash). Six stainless steel washbacks. Wash charger since 2001. Three plain wash stills (10,500 litres charge per still); four lamp-glass spirit stills (one charge at 9,400 litres and three at 4,700 litres). All indirect fired by steam. All shell-and-tube condensers.

MATURATION: Primarily first-fill ex-bourbon barrels from Maker's Mark Distillery; in the past a mix of American and European casks, and in recent years regular bottlings from quarter casks. Eight dunnage and racked warehouses on site with 55,000 casks in total. Five thousand casks at Ardbeg; 40% of production tankered to be matured in Glasgow; 65–70% of the current output of 2.6m L.P.A. is bottled as single malt.

STYLE: Sweet, spicy and peaty.

MATURE CHARACTER: Laphroaig positions itself as 'the definitive Islay malt' because it epitomises the classic Islay character. It also claims on the label to be 'the most richly flavoured of all Scotch whiskies'. The nose is pungent and smoky (coal smoke), with coal tar soap and iodine. The taste is surprisingly sweet to start, then salty and dry, with billows of tarry smoke and medicinal, seaweed-like flavours. Full-bodied.

Linkwood

REGION
Speyside

ADDRESS
Elgin, Moray

PHONE
01343 547004

WEBSITE
www.malts.com

OWNER
Diageo plc

VISITORS
No

CAPACITY
5.6m L.P.A.

HISTORICAL NOTES: Linkwood was founded in 1821 by Peter Brown, factor of Linkwood Estate and agricultural improver, and commenced operation three years later. After 1842 it was managed for him by James Walker (later James Walker & Company), who had come from Aberlour Distillery. On Peter Brown's death in 1868, the distillery passed to his son, William, who demolished and rebuilt it in 1874. The family business became a limited company as Linkwood-Glenlivet Distillery Company Ltd in 1897, and Innes Cameron, a whisky broker in Elgin, joined the board five years later. He would rise to control the company until his death in 1932, when the distillery was acquired by D.C.L.

Major refurbishment took place in 1963, and in 1971 a new distillery was built next door, with four stills (in the 'Waterloo Street' style) – an indication of the excellence of the make, which is ranked Top Class and is used as a 'top dressing' malt in several famous blends. The new distillery was referred to as Linkwood B and operated in parallel with the original (Linkwood A) until 1985, when the old Linkwood closed. The spirits from A and B were vatted together prior to filling into cask. For six months in 2011 the distillery was closed to install a new mash tun and control system, then in April 2013 the old distillery buildings were demolished to allow the Linkwood B stillhouse to be expanded, incorporating the two remaining stills from Linkwood A and six new washbacks.

CURIOSITIES: In 1936 a new Manager was appointed to Linkwood, Roderick Mackenzie – 'a Gaelic-speaking native of Wester Ross, who for many years supervised its making with unremitting vigilance. No equipment was replaced unless it was essential. Even the spiders' webs were not removed for fear of changing its character.' (Professor R.J.S. McDowell)

Professor McDowell is referring to the

The founder's brother was General Sir George Brown, commander of the Light Brigade during the Crimean War.

refurbishment of 1962/63, where the new stills were exact replicas of the ones they replaced, as was the S.M.D. custom.

RAW MATERIALS: Floor maltings until 1963, now unpeated malt from Burghead. Process water from springs near Millbuies Loch; cooling water from the Burn of Linkwood and the Burn of Bogs.

PLANT: Full-Lauter mash tun (12 tonnes). Eleven larch washbacks. Three plain wash stills (14,500 litres charge); three plain spirit stills (16,000 litres charge). All indirect fired. Shell-and-tube condensers.

MATURATION: Mainly refill American hogsheads, some refill European butts. Two thousand casks matured on site, mainly in the Central Belt.

STYLE: Floral, perfumed.

MATURE CHARACTER: A light Speyside style. Estery, with bubblegum, acetone, lemon sherbet, tea roses – clean, fresh and sweet. The taste also is sweet, with traces of white wine (Gewürztraminer grapes), lemon zest and sherbet. Light-bodied.

Littlemill Demolished

REGION
Lowland
ADDRESS
Dumbarton Road,
Bowling, Dunbartonshire
LAST OWNER
Loch Lomond Distillery
Company
CLOSED
1984

HISTORICAL NOTES: Littlemill claimed to have been founded at least by 1772 (when accommodation for Excise officers was built), but before that there was a brewery on the site, attached to Dunglass Castle nearby, which may date from the fourteenth century.

Having passed through several hands, it was improved and enlarged in 1875 by one William Hay, but continued to change hands, until it was bought in 1931 by Duncan Thomas, an American gentleman, who lived in the former exciseman's house. He would go on to build Loch Lomond Distillery in 1965 (see entry), in partnership with Barton Brands of Chicago, agents in the U.S.A. for The Glenlivet until 1971 when the company bought out Duncan Thomas's interests in both Littlemill and Loch Lomond.

Both distilleries were mothballed in 1984, and Loch Lomond was sold to Glen Catrine Bonded Warehouse Ltd the following year.

Barton Distilling (Scotland) Ltd, as the company was now named, came under the control of Amalgamated Distilled Products of Glasgow (part of the Argyll Group), which sold its whisky interests in 1987 to Gibson International. Littlemill resumed production in 1989, when the company also acquired Glen Scotia Distillery in Campbeltown. Two years later, Gibson International and its distilleries were acquired by the company's Managing Director, Ian Lockwood, and Finance Director, Bob Murdoch, in a 'multi-million pound' management buy-out. The restructured company went into receivership in 1994 and its assets were bought by Glen Catrine Bonded Warehouse, who dismantled Littlemill two years later.

Over the following years the buildings deteriorated, then, in 2003, there was a fire. The site has largely been cleared. The surviving stock of Littlemill is now owned by the Loch Lomond Group, who released a 25-year-old in 2015.

Littlemill was a contender for the 'oldest distillery in Scotland' award.

CURIOSITIES: The first thorough government survey of whisky in 1821 showed that Littlemill was then producing 20,000 proof gallons (over 90,000 litres) per annum. By the time Alfred Barnard visited in the mid 1880s it was producing 150,000 gallons, and being sent to 'England, Ireland, India and the Colonies'.

'The locality abounds in charming landscapes not unlike Richmond upon the River Thames. The quiet beauty of the hill slopes and wooded plantations, the hedges covered with summer roses, and the numerous mountain rills, has made this place the favourite resort of artists.'

The distillery once produced 'Dumbuck' (heavily peated) and 'Dunglass' (lightly peated) styles for blending.

Loch Lomond

REGION
Highland (West)

ADDRESS
Alexandria,
Dunbartonshire

PHONE
01389 752781

WEBSITE
www.lochlomond
whiskies.com

OWNER
Loch Lomond Group Ltd

VISITORS
No

CAPACITY
5m L.P.A. malt,
18m L.P.A. grain

HISTORICAL NOTES: Loch Lomond Distillery is a large and practical site, a production unit whose design owes little to aesthetics, without arrangements for visitors. It is situated in an industrial estate on the edge of Alexandria, about a mile and a half from Loch Lomond itself.

The distillery was constructed in 1965/66 by converting a former dye works belonging to the once famous United Turkey Red Company. The job was done by American-born Duncan Thomas (owner of Littlemill Distillery), in partnership with Barton Brands of Chicago, agents in the U.S.A. for The Glenlivet until 1971 when the company bought out Duncan Thomas's interests in both Littlemill and Loch Lomond.

At this time Loch Lomond was producing two styles, Rossdhu and Inchmurrin, from a single pair of stills with rectifying heads (see below). The first dark-grains plant in any malt distillery was also installed here when the buildings were converted.

Like many distilleries, Loch Lomond was mothballed in 1984, then sold in 1985 to Inver House Distilleries, who sold it on the following year to Glen Catrine Bonded Warehouse Company Ltd. This company was the bottling subsidiary of A. Bulloch & Company, a family-owned drinks wholesaler (and at that time retailer as well), in order to secure supplies of malt whisky. In 1992 a second pair of pot stills, copying the first pair, was added, fitted with rectifying heads; in 1994, a Coffey still, to make grain whisky. A third pair of traditional pot stills, with narrow necks, was installed in 1998; and a fourth pair in 2007/08, at which time a unique modified Coffey still, to make malt whisky, was also added.

In March 2014, Loch Lomond Distillers Ltd was bought by a group of senior managers with the support of a private equity company, Exponent, for £210 million. The deal includes Glen Scotia Distillery (see entry), Glen Catrine Bonded Warehouse bottling plant and warehousing in Ayrshire, and

'Loch Lomond' was Captain Haddock's favourite whisky (see Hergé's Adventures of Tintin), but since these books were written between 1929 and 1958, his dram would not have been this Loch Lomond! The owners of the distillery have painted a tanker in the same style as that of Hergé's book – bright yellow with bold black lettering.

a clutch of brands. The new owners are investing over £1 million in the business, and have installed two new stills and three new washbacks at Loch Lomond Distillery.

CURIOSITIES: The range of stills at Loch Lomond, combined with different peating levels, allows it to produce eight different styles of malt whisky and a grain whisky, and makes it almost self-sufficient as a blender. Most of the whisky made here goes for blending. Currently, only Girvan/Ailsa Bay (see pp. 172 and 61) produces both malt and grain whisky on the same site.

Loch Lomond's unusual stills with rectifying heads were inspired by similar plant at Littlemill Distillery. The heads are fitted with perforated plates (as in a patent still), and spirit may be made at different strengths, and different degrees of lightness, up to around 85% A.B.V. (rather than the 70% in traditional pot still distillation).

Spirit from the new modified Coffey still may be drawn off at different strengths from plates all the way down the column. It is much more controllable and consistent than the modified pot stills.

Loch Lomond is the largest loch in the U.K. It is 24 miles long, five miles wide, up to 600 feet deep and has 38 islands. It has been a world-famous beauty spot since the eighteenth century.

The famous song about the Bonnie Banks of Loch Lomond, whose refrain is

Oh ye'll tak' the high road and I'll tak' the low road,
An' I'll be in Scotland before ye',
But wae is my heart until we meet again
On the bonnie, bonnie banks o' Loch Lomond

commemorates two Jacobite prisoners held at Carlisle after the Jacobite Rising of 1745, one of whom was to be hanged and the other released to take the low road home.

RAW MATERIALS: Malt and wheat from independent merchants. Water from bore-holes and Loch Lomond.

PLANT: Full-Lauter mash tun. Thirteen 25,000 litre washbacks and eight 50,000 litre washbacks, all stainless steel. Six traditional pot stills and four Lomond stills, with rectifying columns in place of swan-necks, producing a lighter style of spirit; also a modified Coffey still (uniquely used for making malt whisky. For grain whisky production there are an additional twelve 100,000 litre washbacks and eight 200,000 litre washbacks, and a further Coffey still).

MATURATION: Mainly American oak ex-bourbon casks.

STYLE: The distillery's website states: 'We produce a full range of malts from heavily peated (typical of Islay), to complex fruity (typical of Speyside), to full-bodied fruity (typical of Highland), and also soft and fruity (typical of Lowland).' They are generically bottled under the 'Distillery Select' label, and under the brand names Loch Lomond, Inchmurrin (designed to mature early), Croftengea, Inchmoan, Inchfad, Craiglodge and Glen Douglas. Confusingly, the brand names do not follow the style, 'peated', 'heavily peated' etc. or wood finishes.

Lo

Lochside Demolished

REGION
Highland (East)
ADDRESS
Montrose, Angus
LAST OWNER
Allied Distillers
CLOSED
1992

HISTORICAL NOTES: Lochside Distillery was established within a former Deuchar's brewery in 1957 in the pretty Angus port of Montrose. The brewery dated from 1781, but was rebuilt in the late nineteenth century and was 'far too big for a malt distillery'. Nevertheless, the bold Joseph Hobbs, owner of Ben Nevis Distillery, trading as MacNab Distilleries Ltd, was behind the conversion (see *Ben Nevis*).

His first intention was to make grain whisky, but when other larger grain distilleries were built (notably Invergordon in 1959) he installed four pot stills as well (in 1961), as he had done at Ben Nevis. As at Ben Nevis, he also experimented with the idea of 'blending at birth' – mixing grain and malt new-make prior to filling casks. This experiment was discontinued after his death in 1964, but blending and bottling continued to be done on site, and at one time the blend Sandy MacNab's commanded a loyal following.

In November 1973, ownership passed to Destilerias y Crainza del Whisky S.A. of Madrid (D.Y.C. – pronounced 'Deek'), the first time a European company had bought into the Scotch whisky industry.

The new owner ceased grain production soon after, but continued to produce malt whisky until 1992, originally at the rate of 2.5 million L.P.A. (later cut back by 60%), and exported it in bulk to Spain. In its heyday 35 staff were employed. When D.Y.C. was taken over by Allied Domecq in 1992, production ceased immediately and when mature stocks were depleted by 1996 Lochside was closed and dismantled the following year. The bonded warehouse was demolished in 1999 and the rest demolished in March 2005, following a fire earlier that year which brought down part of the roof. Planning permission for residential housing was immediately applied for.

CURIOSITIES: Deuchar's acquired the site in 1830 and later employed Charles Doig to rebuild it in a 'Continental' style with a tall tower resembling a castle keep, but with a pitched roof. Beer from here was shipped direct to Newcastle by coastal barge.

Longmorn

REGION
Speyside

ADDRESS
Longmorn, by Elgin, Moray

PHONE
01343 551439

WEBSITE
No

OWNER
Chivas Brothers

VISITORS
By appointment

CAPACITY
4.5m L.P.A.

HISTORICAL NOTES: John Duff (who built Glenlossie Distillery in 1876, a quarter of a mile away, whose family owned the lands of Miltonduff, directly to the west, over the River Spey, and who would found BenRiach Distillery next door, two years later) founded Longmorn Distillery in 1893, in partnership with a couple of local businessmen. The location was close to both a good water source and the railway; it had previously been the site of a meal mill founded in the sixteenth century, and long before that of an ancient chapel.

It lasted a mere five years. With the collapse of Pattison's of Leith and a dramatic downturn in the industry, John Duff was obliged to relinquish control of both Longmorn and BenRiach, and by 1901 the Board of Directors included J.A. Dewar (of John Dewar & Sons), Arthur Sanderson (of VAT 69) and James Anderson (of J.G. Thompson, Leith).

In 1899 management passed to James R. Grant, and later his two sons, who became known as 'The Longmorn Grants'; in 1970 they amalgamated with The Glenlivet Grants and the Grants of Glen Grant, and with Hill Thompson & Company, to form The Glenlivet Distilleries Ltd, which was bought by Seagram's in 1977.

The distillery was extended to six stills in 1972 and to eight in 1974, but its appearance has changed little over the years. Seagram was bought by Groupe Pernod Ricard in 2001, and Longmorn is now operated by its subsidiary, Chivas Brothers.

They built a new mash house and tun-room at Longmorn in 2013, equipped with a new Lauter tun and two additional washbacks, increasing capacity from 3.5 to 4.5 million L.P.A.

CURIOSITIES: Until 1980 a railway line connected Longmorn with BenRiach, and latterly a 'puggie' diesel locomotive carried malt from the latter to the former. This engine is now preserved at Aviemore.

The name derives from a seventh-century

The local newspaper began its report on the opening: 'Still another distillery! Evidently the latest one announced for Longmorn is not the last that this district will see . . . When is all this going to end?' However, The National Guardian reported in 1897 that the make had 'jumped into favour with buyers from the earliest day on which it was offered'.

British saint, Eran, to whom Longmorn church is dedicated. The church was named Lannmoeran, meaning 'the enclosure of beloved (St) Eran'. Another account derives the name from Saint Marnoch (or Maernog), who died in 625 and whose feast day was celebrated in many Scottish towns, including, presumably, Kilmarnock. A third source claims that the name comes from the Lhanmorgund (the place of the holy man).

Longmorn has been described as 'the master blender's second choice' (the first being his own creation!), and I certainly know one master blender (retired) who would agree with this!

RAW MATERIALS: Production water from springs on Mannoch Hill. Own floor maltings until 1970, now unpeated malt from independent maltsters.

PLANT: Full-Lauter mash tun (eight tonnes). Ten stainless steel washbacks. Four plain wash stills (10,000 litres charge), four plain spirit stills (6,600 litres charge). All direct fired until 1970; wash stills direct fired until 1990s; now all indirect fired by steam pans and coils. All shell-and-tube condensers.

MATURATION: Refill American oak hogsheads and barrels, some European oak butts.

STYLE: Fruity, with body.

MATURE CHARACTER: Longmorn is a rich whisky and benefits from long maturation in ex-sherry casks. Both the 15YO and the 16YO have discernible traces of sherry on the nose, backed by fruit (oranges, also dried figs) and some malt, and an interesting spicy note – cinnamon? nutmeg? The mouthfeel is big and rounded, the taste sweet (fruit, caramel, malt) then drying slightly towards the finish, which is long and somewhat tannic. Medium-bodied.

Macallan

REGION
Speyside

ADDRESS
Easter Elchies,
Craigellachie, Moray

PHONE
01340 871471

WEBSITE
www.themacallan.com

OWNER
The Edrington Group Ltd

VISITORS
Visitor centre opened
2001

CAPACITY
11m L.P.A.

HISTORICAL NOTES: Originally named Elchies Distillery, Macallan took out a licence in 1824 – one of the first on Speyside to do so. It is likely that there was a farm distillery here before this, since the site is close to one of the few crossing points on the Spey used by cattle drovers. Easter Elchies House, which commands the site, dates from 1700. It has been well-restored by the owners of the distillery.

The founder was Alexander Reid. On his death in 1847, the lease passed through a couple of hands until in 1868 it was taken by James Stuart, a corn merchant in Rothes and owner of Mill of Rothes (see *Glenrothes* and *Glen Spey*). He bought the distillery from the Earl of Seafield in 1886, rebuilt it in stone the same year (the earlier distillery was made of wood) and sold it to Roderick Kemp in 1892. Kemp owned Talisker Distillery from 1879 to 1892, and when he fell out with his partner, attempted to buy Glenfiddich, Cardow and Mortlach before he acquired Elchies Distillery, which he rebuilt, changed its name to Macallan-Glenlivet and greatly improved the quality of the make.

On his death in 1909, ownership passed into a family trust for Kemp's descendants, members of the Harbinson and Shiach families, and they managed it until the present owner bought them out in 1996. A new stillhouse was built in 1954/56 with two wash and three spirit stills, all with shell-and-tube condensers (until then they had worm tubs) and in 1965/66 seven further stills were added in a separate, purpose-built stillhouse, doubling capacity to 90,000 gallons per annum. In 1967 Macallan-Glenlivet became a publicly quoted company, with Kemp's descendants holding 62.5% of the shares, in order to finance the building programme and lay down stock.

The second stillhouse closed in 1990, but was recommissioned and resumed operation in September 2008.

Macallan produced one million gallons for the first time in 1970, but 93% of that went for blending. Two years later sales of bottled Macallan doubled; the board decided to conserve mature stocks and increased production by 24%. The number of stills was again increased in 1974 (to 18) and 1975 (to 21), but that year there was a dramatic drop in orders from blenders – 'the worst downturn in the distillery's history' – encouraging the Directors to allocate further stock for bottling as a single malt.

In 1996 Highland Distilleries (now Edrington) combined its shareholding with that of Suntory to successfully mount a hostile takeover of the company. Edrington took Macallan out of public ownership in 2001.

Since 2011, Macallan's output has increased from eight million to ten million L.P.A. per annum; it increased again by a million litres in the summer of 2014, with the installation of extra washbacks. Eight new warehouses, each holding 26,000 butts of spirit, have been built on the site, and it is planned to increase this to fourteen. This expansion is connected with the ongoing construction of a new distillery capable of producing 17m L.P.A. per annum – Scotland's largest malt distillery. If all goes to plan, this will be completed by March 2017. The two existing distilleries on site will be mothballed.

CURIOSITIES: In the late nineteenth century, Easter Elchies House was rented by the Earl of Elgin as a shooting lodge. It was while on holiday here that the ninth Earl received word that he had been created Viceroy of India.

The Macallan has frequently commanded the highest prices at auction of any malt whisky (see also *Dalmore*). In January 2014, a six-litre bottle – one of only four, made by Lalique – was sold by Sotheby's in Hong Kong for $HK 628,000, the proceeds going to charity. One of the remaining

Macallan has the smallest spirit stills on Speyside (illustrated on Bank of Scotland £10 notes), and the largest single-roofed warehouse in Europe.

bottles was sold privately and the other two are in the distillery's archive.

RAW MATERIALS: Soft water from bore-hole aquifers on site. Floor maltings closed late 1950s. Unpeated malt mainly from Simpson's of Berwick-upon-Tweed.

PLANT: *Stillhouse 1:* Full-Lauter stainless steel mash tun with a peaked canopy (7.5 tonnes per mash), 16 stainless steel washbacks. Five plain wash stills (12,750 litres charge), ten plain spirit stills (3,900 litres charge; each wash still charges two spirit stills). Wash stills direct fired by gas; spirit stills indirect fired by steam coils. Shell-and-tube condensers.

Stillhouse 2: New full-Lauter mash tun, with a deep bed (six tonnes per mash), six new Douglas fir washbacks. Two plain wash stills (12,000 litres charge; bases replaced, heads and lyne arms original), four plain spirit stills (3,900 litres charge, all original). All indirect fired. Shell-and-tube condensers (five original, one replaced).

MATURATION: Dry-Oloroso sherry, European and American oak first and refill butts and hogsheads for traditional Macallan, with some refill bourbon barrels and American oak sherry-seasoned casks for some bottlings.

STYLE: Rich, robust, oily, and fruity.

MATURE CHARACTER: The Macallan is the benchmark 'sherried' style of malt. The nose is rich and chocolatey, with dried orange peel, dried fruits, sherry, nuts and a trace of sulphur. The mouthfeel is smooth and voluptuous; the taste sweetish to start, but tannic-dry overall, with sherry, Christmas cake (burnt edges), caramel. Full-bodied.

Macduff

REGION
Highland (East)

ADDRESS
Macduff, Moray

PHONE
01261 812612

WEBSITE
No

OWNER
John Dewar & Sons Ltd
(Bacardi)

VISITORS
By appointment

CAPACITY
3.34m L.P.A.

HISTORICAL NOTES: Macduff Distillery was a child of the Sixties. It was built in 1960 by a consortium which included Brodie Hepburn (whisky blenders in Glasgow, with interests in the recently built Tullibardine and Deanston Distilleries), trading as Glen Deveron Distillers Ltd. It was designed by William Delmé-Evans (see *Tullibardine, Jura*) and incorporated several novelties which have now become commonplace, such as indirect firing by steam coils, shell-and-tube condensers and a stainless steel mash tun.

Glen Deveron Distillers sold to Block, Grey & Block in 1966, who added one still that year and another in 1968, and sold to William Lawson Distillers in 1972, a subsidiary of the Italian company Martini & Rossi. The tun room and stillhouse was rebuilt in 1990, and a fifth still installed. Bacardi bought Martini & Rossi, including William Lawson Ltd, Macduff Distillery and the Glen Deveron brand in 1992, and ownership was transferred to John Dewar & Sons when Bacardi acquired that company from Diageo in 1998.

Stills generally work in pairs, and so an even number of stills is the norm. Only two distilleries have five stills, Macduff and Talisker.

CURIOSITIES: The model village of Macduff was laid out by James Duff, second Earl of Fife, in 1783, planned around one of the best harbours on the Moray Firth, which became a leading herring port during the nineteenth century, curing and exporting fish direct to the Baltic. There was great rivalry between Macduff and the older and more gracious Royal Burgh of Banff, a mile away across the Deveron Estuary.

William Lawson was Manager of the Irish wine & spirits merchants E. & J. Burke of Dublin in 1889, who sold their Scotch under the Lawson's label. This trademark was bought by Martini & Rossi in 1963 and amended to avoid any confusion with the D.C.L. brand Peter Dawson. In 1969 all Martini & Rossi's whisky interests were placed under the

Unknown in the U.K., William Lawson's blended Scotch sells around 15 million bottles annually, especially in southern Europe and Mexico. Almost all the make from Macduff goes into this and other blends.

newly created William Lawson Distillers Ltd. At the same time, the D.C.L. claimed they owned the brand name Macduff, so the malt was named Glen Deveron.

In the mid 1970s Glen Deveron 5YO was number three malt in the world, on account of its popularity in Italy and France. In April 2013, Dewar's introduced a range of three expressions of Glen Deveron for the duty free market, at 16, 20 and 30YO, named the 'Royal Burgh Collection', and in 2015 introduced a new core range, now named The Deveron and filled into aquamarine bottles, at 10, 12 and 18YO.

RAW MATERIALS: Unpeated malt from independent maltsters. Soft process water from bore-holes on site and springs. Cooling water from the Gelly Burn.

PLANT: Full-Lauter mash tun (10 tonnes). Nine stainless steel washbacks. Two plain wash stills (15,000 litres charge), three plain spirit stills (16,200 litres each charge). All indirect fired by steam pans and coils. Shell-and-tube condensers; spirit stills mounted horizontally, with after-coolers.

MATURATION: No filling on site since 2002; tankered to Coatbridge, matured in Central Belt. Mix of first-fill and refill hogsheads and butts.

STYLE: Malty, nutty-spicy.

MATURE CHARACTER: Medium-rich and sweet on the nose, with malty and sherry notes. The taste is sweet, with apples and pears, and perhaps mangos and papayas, then drying into nuts and cereals. Medium length; medium-bodied.

Mannochmore

REGION
Speyside

ADDRESS
Birnie, Elgin, Moray

PHONE
01343 860331

WEBSITE
www.malts.com

OWNER
Diageo plc

VISITORS
No

CAPACITY
6m L.P.A.

Bottles of Loch Dhu now fetch around £300 at auction.

HISTORICAL NOTES: Mannochmore is another example of Scottish Malt Distillers' building frenzy during the 1960s/early 1970s. The new distillery was raised in 1971 adjacent to the company's Glenlossie Distillery, which itself had been tripled in capacity ten years previously. Like several of the distilleries built during this period, in what became known as the 'Waterloo Street' style, efficiency was everything (see *Caol Ila*). Its original purpose was to provide fillings for the Haig blends, which had been market leaders in the U.K. during the 1950s/1960s, but which were now in decline.

In 1996 a famous expression of Mannochmore was released under the name 'Loch Dhu – the Black Whisky'. It was heavily tinted with spirit caramel (so much so as to have a bitter taste), and was designed to be mixed with Coca-Cola or ginger ale. It enjoyed popularity in Denmark, but nowhere else, and was soon discontinued. In 2013 capacity was increased by the addition of another pair of stills and a state-of-the-art Briggs mash tun installed.

RAW MATERIALS: Lightly peated malt from Burghead. Water from the Mannoch Hills via the Barden Burn.

PLANT: Briggs full-Lauter mash tun (11.5 tonnes). Eight larch washbacks, eight stainless steel washbacks. Four plain wash stills (14,400 litres), four plain spirit stills (17,000 litres charge). All indirect fired. Shell-and-tube condensers.

MATURATION: Refill American hogsheads. Matured mainly in the Central Belt.

STYLE: Fruity.

MATURE CHARACTER: The nose is light and fruity, fresh and bright, with breakfast cereals. The taste is Speyside-sweet, with a creamy mouthfeel, fresh fruits and a woody dryness in the short finish. Light-bodied.

Millburn Redeveloped

REGION
Highland (North)
ADDRESS
Millburn Road, Inverness
LAST OWNER
D.C.L./S.M.D.
CLOSED
1985

The distillery ceased production in 1985 and by 1989 was a Beefeater steak house.

HISTORICAL NOTES: It is not known for certain when Millburn was established, possibly in 1807, though the first written record was in 1825. The site is about a mile east of the centre of Inverness, adjacent to the former Cameron Highlanders barracks.

Production ceased in 1837. In 1853 the feu was taken by a local corn merchant, David Rose, who used the building as a flour mill – there were five mills drawing water from the Mill Burn at that time. In 1876 he applied to use the town water supply and commissioned a new and larger distillery on the site.

The property went to David Rose's son, George, in 1883 and he sold to Alexander Price Haig and his brother David, of the famous distilling dynasty, in 1892. They refurbished: 'the whole of the internal arrangements have been remodelled and the plant and machinery are entirely new'. But the depression in the industry following World War I persuaded them to sell to Booth's Distillers Ltd, the famous gin makers, in 1921, for £25,000. In April the following year fire destroyed most of the distillery buildings and large stocks of barley and malt, the cost estimated at £40,000. The fire brigade was 'greatly assisted' by men of the Cameron Highlanders in controlling the fire: Lt Col. David Price Haig had been a territorial officer in the regiment for 30 years.

Reconstruction was entrusted to Charles Doig of Elgin, and the new distillery, which opened in 1887, was capable of producing 150,000 gallons a year – nearly twice the former output.

Booth's bought Wm Sanderson & Son Ltd (blenders of VAT 69) in 1935, and was itself taken over by D.C.L. in 1937, and managed by S.M.D. from 1943. Production ceased during World War II. In 1958 mechanical stoking was installed, and the two stills were converted to indirect firing by steam coils in 1966. Saladin box maltings had been installed two years before, when electric power was also introduced.

CURIOSITIES: Booth's Distillery Ltd, the famous gin distillers, used the make for their house whisky, Cabinet, and later for VAT 69. Under S.M.D., Millburn was licensed to Macleay Duff, and went into their blends, including a blended malt at 12YO.

The buildings that once housed the distillery now house the Auld Distillery Restaurant and Pub, which has some interesting whisky memorabilia.

Mi

Miltonduff

REGION
Speyside

ADDRESS
Miltonduff, Elgin, Moray

PHONE
01343 547433

WEBSITE
No

OWNER
Chivas Brothers

VISITORS
By appointment

CAPACITY
5.8m L.P.A.

HISTORICAL NOTES: The distillery was built six miles south-west of Elgin, in the grounds of Pluscarden Abbey, on account of the quality of the water source.

Not surprisingly, many illicit distillers used the vicinity in the late eighteenth century – around 50, it is claimed – and the present distillery, founded in 1824, is on the site of one of them, Milton, where once stood the abbey's meal mill. The name was changed when the Duff family bought the land, hence Milton of Duff. It retains an old water wheel.

It was acquired by William Stuart (a co-proprietor of Highland Park Distillery) in 1866; he was joined by Thomas Yool in 1890 and the distillery was extended in the mid 1890s, at which time it was producing over one million litres of alcohol.

In 1936 Yool sold to Hiram Walker, who had bought George Ballantine & Company the year before. In 1974/75 'modernisation visited Miltonduff with a vengeance', raising capacity to 5.24 million litres per annum – one of the largest distilleries in Scotland. A reception centre was also installed at this time, but it soon closed.

In 1986 Allied Distillers acquired 51% of Hiram Walker, and the rest the following year. In 2005, the majority of Allied Distillers' whisky interests were bought by Pernod Ricard, including Ballantine's and Miltonduff Distillery, which is now operated by their subsidiary Chivas Brothers.

CURIOSITIES: Pluscarden was originally a priory, endowed by King Alexander II in 1230. In 1454 it absorbed the old Benedictine Priory of Urquhart, only to fall into disuse at the Reformation. Nearly 400 years later it was gifted to the Benedictine Community of Prinknash by Lord Colum Crichton-Stuart; monks again took up residence in 1948, while the buildings were being restored. In 1974 it was elevated to the status of Abbey.

Between 1964 and 1981, two Lomond stills were in operation at Miltonduff, producing Mosstowie

Pluscarden Abbey was once famous for its ale – so good that 'it filled the abbey with unutterable bliss'. The excellence of this brew was attributed to the Black Burn, which had been blessed by a saintly abbot in the fifteenth century, and which supplies the distillery today.

single malt (see *Dumbarton, Inverleven*). The Lomond style of still was invented in 1955 by Hiram Walker's leading chemical engineer, Alistair Cunningham. It had an unusually wide, drum-like neck, which held three rectifying plates, similar to those found in a column still. The purpose of the plates was to vary the amount of reflux, and thus the styles of whisky produced. They could be rotated – the horizontal position would maximise reflux, and the vertical position minimise it. They could be used dry or filled with water, which, again, increased reflux. Mosstowie was lighter in character than Miltonduff.

Miltonduff also housed Allied Distillers' Malt Distilleries Technical Centre, with laboratory, engineering department and warehouse management offices. Chivas Brothers' Northern Operations Headquarters is now based there. The laboratory was moved into Glen Keith Technical Centre. CO_2 extraction from the washbacks was pioneered here.

In the 1960s a novel method of heating the wash stills was pioneered at Miltonduff. The wash passed through a series of heat exchangers (using hot water from the condensers), which would heat it to 75–80° prior to charging the still. Once the still was charged, the wash was drawn through another heat exchanger, where steam heated it to boiling point. It was returned to the still as vapour via a diffuser, and would in turn heat the residual wash in the still. This continued until distillation was complete.

RAW MATERIALS: Unpeated malt from Kilgours, Kirkcaldy; own floor maltings removed in the 1970s. Process water from a spring on site; cooling water from the Black Burn.

PLANT: Full-Lauter mash tun, with copper dome (14.5 tonnes); 16 stainless steel washbacks. Three plain wash stills (18,100 litres charge); three plain spirit stills (18,400 litres charge), all dumpy, with broad necks and sharply descending lye arms. All indirect fired. Shell-and-tube condensers.

MATURATION: Mainly refill U.S. hogsheads, some first-fill.

STYLE: Sweet, grassy and fragrant, with some spice.

MATURE CHARACTER: Miltonduff has always been a blending malt, and the few expressions that have appeared as singles reflect this. The nose is grainy and malty, with some honey notes and light floral scents. The taste is sweet, with some fruit and nuts. Medium-bodied.

Moffat Malt & Grain Distillery (Dismantled)

ADDRESS
Airdrie, Lanarkshire
OWNER
Inver House Distillers
CAPACITY
30m L.P.A.
CLOSED
1986

It is said that the name Killyloch should have been Lillyloch (from where the site draws its process water), but the stencil had been wrongly cut.

HISTORICAL NOTES: The distillery complex (known as Moffat after the mills which once stood there) is on the outskirts of Airdrie, and included a spread of 30 high-racked maturation warehouses (holding around 480,000 casks), a bottling plant and offices. The distillery itself was removed in the 1980s.

The location was chosen on account of its excellent water supply, the ready availability of labour and its geographical centrality, with good road links. There had been a paper mill here, which had been closed for about three years when Inver House acquired the somewhat derelict site.

Inver House was at this time a subsidiary of Publicker Industries Inc. of Philadelphia, and the intention was to produce both malt and grain whisky on the same site, principally to meet the demand for Inver House Green Plaid, a blended Scotch, in the U.S.A. Conversion of the old mill began in the spring of 1964; the first malt whisky was distilled here in February 1965, and the first grain whisky a month later. Three different styles of malt whisky were made in three pairs of stills, named Killyloch, Glenflagler and Islebrae, with varying degrees of peating (Killyloch unpeated, Glenflagler lightly peated, Islebrae heavily peated). Islebrae was originally named Glen Moffat. The grain whisky from Moffat was named Garnheath.

Malt whisky distilling ceased in 1982; grain distilling in 1986. In 1988 Inver House was bought from its American owners by its management (led by Bill Robison) for £8.2 million. The company had made a loss of £2.12 million the year before, taking accumulated losses to £9.7 million. It has since been a resounding success, the directors selling to Pacific Spirits, a Thai company, for £56 million in 2001.

CURIOSITIES: The size of the Moffat complex at the time of its construction dictated that it have its own maltings, so the Wanderhaufen system of box maltings (known as the 'moving street') was installed, soon after the distillery itself had gone into production – first one 'street' and ultimately seven, steeping up to 700 tonnes of barley a day, with four kilns. At this time it was the largest maltings in Europe. The maltings closed in 1978.

At its peak in the 1970s a thousand people were employed in the Moffat distillery complex.

The grain whisky was never bottled. The four malts have appeared but only very occasionally, from independent bottlers. Glenflagler was bottled by Inver House in the mid 1980s, at five and eight years old.

Mortlach

REGION
Speyside

ADDRESS
Dufftown, Moray

PHONE
01340 822100

WEBSITE
www.malts.com

OWNER
Diageo plc

VISITORS
By appointment

CAPACITY
3.8m L.P.A.

Mortlach was the first distillery to be built in Dufftown, which would in time become a major distilling centre with nine distilleries, six of which are still in operation.

HISTORICAL NOTES: Mortlach was first licensed in 1823 to James Findlater, who used a site that had been formerly used by smugglers on account of the excellent water from Highlander John's Well. He ran it in partnership with two local men, Alexander Gordon and Donald McIntosh, but in 1831 it was sold for £270 to John Robertson. By 1842 it was owned by J. & J. Grant, who were building Glen Grant Distillery at the time and removed the distillery plant to their new site in Rothes. The granary at Mortlach was used as a Free Church, until one was built in Dufftown.

In the late 1840s the buildings were bought by John Gordon, who initially installed a brewery there before resuming distilling, after 'enlarging and improving the works'. He sold his whisky 'mostly in Leith and Glasgow, but [it has] attained some celebrity in the district under the name of *The Real John Gordon*'. (*Elgin Courant*, 1862)

He resumed distilling in 1851 and two years later took George Cowie into partnership. Cowie had been a railway surveyor, and would become Provost of Dufftown, and by 1869 was outright owner of Mortlach Distillery. His son Dr Alexander Mitchell Cowie, who was a senior medical officer in Hong Kong, returned to join his father in 1895. He became a leading figure in the whisky industry and Deputy Lieutenant of Banffshire: over the next 30 years he built a high reputation for Mortlach (it is ranked Top Class by blenders), doubling the distillery's capacity (to six stills) in 1897, when a private railway spur, shared with Glendullan, was installed.

Dr Cowie's only son was killed in World War I, and in 1923 he sold the distillery to John Walker & Sons, a key customer for fillings, and thus it joined D.C.L. in 1925. The distillery was managed by S.M.D. from 1936. Unlike most malt distilleries, Mortlach stayed open during World War II, except in 1944.

Most of the distillery buildings were demolished and rebuilt in the early 1960s (completed 1964),

Mortlach is a key filling for the Johnnie Walker blends.

introducing mechanical coal-stoking to the stills (they were converted to indirect firing in 1971).

Mortlach is a famously rich malt, ranked Top Class by blenders and very popular with connoisseurs, but since it is a key filling for the Johnnie Walker blends, it was difficult to come by. This changed in 2014, when a core range of three expressions (Rare Old, 18YO and 25YO) was launched globally. Plans were also announced that year to build a replica distillery adjacent to Mortlach, though these were put on hold in December 2014.

CURIOSITIES: Mortlach has possibly the strangest set of stills in the business, and the most complex distilling regime. There are six stills, but all different sizes and shapes, and the regime is not quite triple distillation; actually it's 2.8. It takes six months of training for a new stillman to work it out. There is no space here to describe the regime, but crucial to the equation is No. 1 spirit still, known as the 'Wee Witchie', which is charged three times in each run.

'There is not perhaps a distillery in Scotland that has so many private customers as Mortlach from which spirits are sent not only over the three kingdoms to families, but to America, India, China and Australia, in all of which Mr Cowie has customers who prefer his distillation to all others . . .' (*Elgin Courant*, 1893)

Over the 20 years from 1866 to 1886, William Grant of Glenfiddich learned his trade at Mortlach, working in all parts of the business and finishing as Manager, before starting his own enterprises down the road.

RAW MATERIALS: Floor maltings until 1968, now unpeated malt from Burghead. Soft process water from Guidman's Knowe springs; cooling water from springs in the Conval Hills.

PLANT: Full-Lauter tun (12 tonnes). Six larch washbacks. Three boil-ball wash stills (two 7,500 litres charge, one 16,000 litres charge); three boil-ball spirit stills (one 7,000 litres charge [the 'Wee Witchie'], one 9,300 litres charge). Direct-fired until 1971, now all indirect-fired. Worm tubs.

MATURATION: For proprietary bottlings refill European casks are used. Some matured on site, the rest in the Central Belt.

STYLE: Heavy, meaty.

MATURE CHARACTER: A big, rich whisky. Dry fruitcake, with burnt edges, moistened with Madeira. Soft, full-bodied texture; hint of allspice in the taste. Long finish.

North British Grain Distillery

ADDRESS
Gorgie, Edinburgh
PHONE
0131 337 3363
WEBSITE
www.northbritish.co.uk
OWNER
Lothian Distillers Ltd
CAPACITY
73m L.P.A.

The North British is the sole remaining distillery in Edinburgh, which once had many, among them Canonmills, Lochrin, Bonnington, Leith, Sunbury, Abbeyhill and Edinburgh Distilleries. It is currently the third largest distillery in Scotland.

HISTORICAL NOTES: In 1885 a group of independent blenders and spirits merchants banded together to build a distillery in Edinburgh for the supply of grain whisky, in response to the monopoly of D.C.L., with its 'varying qualities and fluctuating prices'. Their goal was 'to check a great monopoly and maintain a uniformly low price'. Andrew Usher II was the first Chairman, John Crabbie Vice-Chairman and William Sanderson Managing Director.

A greenfield site was chosen, to the north of Gorgie Road and west of Dalry (and the Caledonian Distillery). It was well-connected by railway lines (notably the Wester Dalry branch of the Caledonian Railway, connecting to Glasgow, and the Granton & Leith branch of the same railway, running to Leith, principal grain port in Scotland), and with ample water from the Union Canal, which passed close by.

Production commenced in September 1887, with one Coffey still, and in the first year 1.5 million gallons were made and sold; within four years capacity was doubled, so by 1897 production had risen to three million gallons, and the whole of the next year's production had been sold in advance. The Chairman announced, 'there is no whisky more popular in Scotland'.

Difficult trading during the 1920s led to an arrangement with D.C.L. to ration their annual production to the amounts ordered for the ensuing year. Since D.C.L. usually sold about four times as much spirit as North British, production was divided four-fifths/one-fifth, reducing North British output to 2/2.5m gallons a year. This arrangement lasted until 1934, by which time Prohibition had been repealed in the U.S. and sales rose sharply. Then came World War II and such a shortage of grain that the North British had to cease distilling in 1939.

Although distilling resumed at the end of hostilities, grains were rationed until 1949. The 1950s were years of expansion and modernisation.

During World War I the North British was required to make spirits for munitions, and was then closed altogether as grain supplies dwindled.

An extensive site – a former tram depot – was acquired to the west of the distillery, and the former Scottish Brewers site on Slateford Road, to the south of the distillery. A new site was developed at Muirhall, 18 miles away, beyond West Calder, capable of holding 40m proof gallons in 26 warehouses. Production capacity increased even more rapidly to meet orders which were climbing year on year. In 1955, output stood at 3m gallons; by 1961 it was 6m gallons, and by 1972, 12m gallons. At this time the company was employing around 400 people.

In 1968 an evaporation plant for dreg was installed, and a dark-grains plant producing pelletised cattle feed. This was extended in 1976 to process draff and dreg from Caledonian Distillery nearby. CO_2 was pumped to Caledonian for recovery.

In 1993, management of the North British was assumed by Lothian Distillers Ltd, a partnership between I.D.V. and Robertson & Baxter. I.D.V. merged with U.D. in 1998 to become U.D.V. and later Diageo. R. & B. changed its name to Edrington in 1999.

CURIOSITIES: In 1947 the North British's transport fleet consisted of a horse and cart; in 1967 it boasted six Commer motor lorries (which delivered casks as far away as Aberdeen) and four bulk tankers. It also owned two diesel locomotives to pull the 'grain train' to and from the Port of Leith.

In 1959 W.G. Farquharson, Chairman of North British and of Arthur Bell & Sons, proposed building a Lowland malt distillery within the complex (as other distillers had done), but this came to nothing.

The 250 millionth cask of North British proof grain whisky was distilled on 12 November 1970. It was sealed in a commemorative keg and is displayed at the distillery.

RAW MATERIALS: Maize mainly from the South of France. Water from reservoirs in the Pentland Hills. Drum maltings until 1948, then Saladin boxes until 2007, now green malt from independent maltsters. Note: it is the only remaining distillery to use green malt. Mixed in proportion 1:4 – the highest malt content of any grain whisky.

PLANT: Two mash tuns. Three Coffey stills. Patent still with three columns decommissioned in 2007, but still in situ.

MATURATION: Mainly U.S. ex-bourbon first-fill and refill casks; filled and matured at Muirhall, West Calder.

STYLE: Robust. More meaty than other Scotch grain spirits.

North Port Demolished

REGION
Highland (East)
ADDRESS
Brechin, Angus
LAST OWNER
D.C.L./S.M.D.
CLOSED
1983

North Port was closed in May 1983 due to the world recession at that time and as part of S.M.D.'s policy to reduce output to bring the level of maturing stock in line with anticipated future sales.

HISTORICAL NOTES: A 'port' is the old Scots word for a town gate, but by the time the distillery was built in 1820, Brechin's medieval North Port had long disappeared. The town itself is of ancient foundation, the cathedral here having been endowed by King David I in 1150. It stands in the midst of excellent corn country and local farmers carted their crop into the malting barns: the founder's family had farmed in the district for two generations.

The distillery was built by a local worthy, David Guthrie, who had established the town's first bank in 1809 and served as Provost. Its original name was the Townhead Distillery Company, but in 1823 this was changed to Brechin Distillery Company.

Ownership passed to the founder's sons, and a limited company was established in 1893. In 1922 it was bought by D.C.L. in partnership with W.H. Holt & Company (wine and spirits merchants in Manchester) and wound up. The distillery and stocks were sold to S.M.D. The make was used for blending, except a small amount in the early 1980s, which was bottled as a single by D.C.L. subsidiary John Hopkins & Company, for sale in Italy.

CURIOSITIES: The *History of Brechin* (1839) reports: 'Formerly the neighbourhood of Brechin was much infested with bands of smugglers, carrying whisky from the Grampian Highlands to the low country, and Brechin itself depended upon these *merchants* for the supply of *mountain dew*. Now the matter is reversed. There is an extensive distillery in the town, called the North Port Distillery [supplying] a far purer spirit than was formerly drunk under the name of smuggled whisky . . . [and whisky was] the chief potation of all classes [in Brechin].'

North of Scotland

Grain Distillery (Demolished)

ADDRESS
Cambus, Alloa,
Clackmannanshire

OWNER
North of Scotland
Distilling Company Ltd

CLOSED
1980

HISTORICAL NOTES: Strathmore Distillery was established as a private venture by George Christie (see *Strathspey*) in 1957, on a 1.5-acre site formerly used by Robert Knox's Forth Brewery (established 1786) at Cambus, close to the distillery of that name in Alloa. Originally, it produced malt whisky in three modified Coffey stills, but by 1960 had switched to grain whisky production – the smallest grain whisky distillery in Scotland.

The distillery was closed in 1980, sold to D.C.L. in 1982 and dismantled by 1993. The buildings were demolished soon after.

CURIOSITIES: The distillery was said to be haunted by a former brewer, whose apparition was repeatedly 'sighted by both employees and Excise officers'. (Phillip Morrice)

'The North of Scotland Distillery keeps a higher than usual proportion of feints and foreshots [congeners] when distilling its whisky in order to give it more character, so that it can be recognised as North of Scotland grain.' (Phillip Morrice)

The distillery's name changed in 1964 to North of Scotland Distillery, although Clackmannanshire is a long way from the 'north of Scotland'.

Oban

REGION
Highland (West)
ADDRESS
Stafford Street, Oban,
Argyll
PHONE
01631 572004
WEBSITE
www.malts.com
OWNER
Diageo plc
VISITORS
Visitor centre and shop
CAPACITY
870,000 L.P.A.

HISTORICAL NOTES: Founded in 1794, Oban is among the earliest surviving malt whisky distilleries. Its founders were John and Hugh Stevenson, local worthies with interests in slate quarrying, house building and shipbuilding in and around Oban since 1778. The town of Oban grew up around the distillery, which is squashed between the main street and a high cliff. The brothers had originally equipped the site as a brewery in 1793, and continued to trade under the name the Oban Brewery Company.

The distillery remained in their family until 1866, and after one other owner it was acquired in 1883 by J. Walter Higgin, who refurbished and modernised it. Three years previously the West Highland railway line, connecting Oban to Glasgow, was opened. The town became a popular tourist resort, and Higgin could send his whisky direct to market in Glasgow.

In 1898 the distillery was bought by a consortium led by the redoubtable Alexander Edward, a well-known figure in the whisky trade (see *Aultmore*), with the support of Buchanan's, Dewar's and Mackie's (White Horse). In 1923 Buchanan-Dewar took over Oban Distillery, and thus it joined D.C.L. in 1925.

The make was chosen by U.D. to represent the West Highland style in their Classic Malts series (1988).

CURIOSITIES: The building which houses the Manager's office was formerly the dwelling of the Stevenson Brothers who founded the distillery. By the time Alfred Barnard visited in 1885, Higgin had converted it into offices, but retained the peephole door in the former sitting room, which the Stevensons had installed to keep an eye on operations in the stillhouse. In 2015 the standard 14-year-old expression was joined by Oban Little Bay: *Oban* is Gaelic for 'Little Bay'.

In the 1890s, when the cliff behind the distillery was being dynamited to make room for a new warehouse, caves were discovered containing human remains and artefacts dating from the Mesolithic era (4,500–3,000 BC).

RAW MATERIALS: Floor maltings until 1968; now lightly peated malt from Roseisle. Soft water from Loch Gleann a' Bearraidh.

PLANT: Traditional mash tun (6.5 tonnes), using a sparge ring. Four European larch washbacks. One lamp-glass wash still (11,600 litres charge); one lamp-glass spirit still (7,000 litres charge). Both indirect fired. Rectangular worm tubs, both worms in same tub (as at Cragganmore).

MATURATION: Refill American hogsheads. Matured mainly in the Central Belt. Around 4,000 casks on site.

STYLE: Fruity, lightly maritime.

MATURE CHARACTER: The seaside location of the distillery somehow communicates itself to the flavour of the whisky, in spite of it being matured inland. The nose is fresh and maritime, with seaweed and salt behind fresh fruits and a hint of smoke behind that. The mouthfeel is soft and slightly oily; the taste is sweet, with dried figs and light spiciness, a trace of salt and a thread of smoke. Medium-bodied.

Parkmore

REGION
Speyside
ADDRESS
Dufftown, Moray
OWNER
The Edrington Group
CLOSED
1931

'Externally the most perfect survivor of the late 1890s boom in distilleries.'
Charles Craig

HISTORICAL NOTES: Parkmore was an elegant small distillery, typical in style to others of the 1890s, designed by Charles Doig, and the fifth distillery to be built in Dufftown. James Watson & Company, a well-known Dundee blending house (established in 1815 and owners of Ord and Pulteney Distilleries, and of the blend Baxter's Barley Bree) was behind the scheme, and took whole ownership in 1900.

In 1923, Watson's was acquired by Buchanan-Dewar's and John Walker & Sons; Dewar's took the distilleries and eight million gallons of mature stock was shared between them. It was described as 'one of the most important stocks of old whisky in the country'. All three (including Watson's) joined D.C.L. in 1925.

Parkmore was managed by S.M.D. from 1930 and was closed in 1931. The site was licensed to Daniel Crawford & Son Ltd, a very minor D.C.L. subsidiary, and used for warehousing and stores. It was sold for the same purpose to Highland Distilleries (now the Edrington Group) in 1988.

Pittyvaich Demolished

REGION
Speyside

ADDRESS
Dufftown, Moray

LAST OWNER
United Distillers

CLOSED
1993

HISTORICAL NOTES: Pittyvaich-Glenlivet was built by Arthur Bell & Sons Ltd in 1975, next door to their Dufftown-Glenlivet Distillery, and operated in parallel with it. A dark-grains plant was built at the same time to look after the needs of both distilleries.

The distilleries were built on the site of Pittyvaich House, which dated from the 1850s and was demolished to make way for Pittyvaich. The new buildings were described as 'ultra-modern, so they bear little resemblance, if any, to the kind of distillery you see along Speyside' by Bell's historian, Jack House (1976). 'They started with the plant and built the outside of the distillery around it.'

Dufftown had been extended to four stills in 1967 (and a further pair would be installed in 1979). Pittyvaich had two pairs of stills – such was the demand for Bell's Extra Special at the time. The new distillery had a capacity of a million proof gallons.

U.K. sales of Bell's rose from £8.8 million in 1970 to £159 million in 1980, by which time the brand was the most popular Scotch in Britain, commanding 25% of the market.

By the early 1990s Pittyvaich's buildings had deteriorated, and its asbestos roofs had to be replaced. It was decided to close the site in 1993 and the distillery was demolished in 2002.

CURIOSITIES: The prefix 'Pit-' is one of the very few words that have come down to us from Pictish, in which times it signified a farmstead, place or landholding. The prefix is common, especially in the Eastern counties, the former kingdoms of the Picts. 'Vaich' is Gaelic and may mean either 'byre' (cattle-shed), possibly 'cattle-field' or perhaps 'birchwood'.

Port Dundas

Grain Distillery (Demolished)

ADDRESS
North Canal, Borron
Street, Glasgow
PHONE
0141 332 2253
LAST OWNER
Diageo plc
CLOSED
2009

HISTORICAL NOTES: There have been three Port Dundas Distilleries at the Glasgow end of the Forth and Clyde Canal. The first was founded in 1811, the second in 1813 and the third in 1838. The last only operated for two years; the first two amalgamated in 1845, when Coffey stills were installed in parallel with their pot stills – Port Dundas continued to make grain whisky in pot stills until at least the late 1880s.

The owner, M. Macfarlane & Company, successor to Daniel Macfarlane, the founder of one of the distilleries, joined D.C.L. on its foundation in 1877 (nee Commonwealth, Carsebridge), at which time the distillery had three Coffey and five pot stills – one of them the largest in the industry. At the time of Alfred Barnard's visit ten years later it was the largest distillery in the British Isles, producing 12 million L.P.A. on a nine-acre site. It was badly damaged by fire in 1903, rebuilt by 1914, with a new drum maltings and situated within one of the first reinforced concrete buildings in Europe (closed and gutted 1983), and was again badly damaged in 1916 when No. 6 warehouse went up in flames, along with 12,000 hogsheads of whisky.

The distillery was transferred to Scottish Grain Distillers in 1966 and extensively modernised during the 1970s, to double output at a cost of £10 million. With the site increasing to 25 acres, following the acquisition of neighbouring maltings and fertiliser plant, a super-efficient dark-grains plant replaced the original in 1977 and a CO_2 recovery plant was also installed (this gas is a by-product of fermentation). At this time Port Dundas was D.C.L.'s flagship distillery. In 1992 the licence was taken by U.D.'s subsidiary United Malt and Grain Distillers.

The distillery was terminally closed in 2009 and the site has been cleared.

Alfred Barnard was impressed by the piggery at Port Dundas, which accommodated 400 pigs, including some 'highly-bred animals of great size' housed in pens decorated with their prize cards. Swine fever wiped them out around 1900, and they were not replaced.

CURIOSITIES: Port Dundas was part of the Forth and Clyde Canal project (completed 1790), and took its name from Sir Lawrence Dundas of Kerse, President of the Canal Company. The distillery was built on the north bank of the canal and made good use of both it and the nearby railway line to carry goods in and out. The canal closed in the 1960s (stretches of it have been reopened in recent years for recreation), and subsequently all materials come by road.

RAW MATERIALS: Wheat and maize (since 1955). Process water from Loch Katrine (the distillery was licensed to draw water from the canal, but it was found to be neither cold enough nor clean enough).

PLANT: Three Coffey stills, one stainless steel.

MATURATION: Mainly American oak, first-fill and refill ex-bourbon hogsheads and barrels.

STYLE: Traditionally full-bodied, and having a poor reputation in the 1960s (Morrice), Port Dundas make has become lighter and cleaner since the improvements of the 1970s.

Port Ellen Dismantled

REGION
Islay

ADDRESS
Port Ellen, Isle of Islay

OWNER
Diageo plc

CLOSED
1983

HISTORICAL NOTES: Port Ellen single malt enjoys cult status among collectors who favour the smoky style. They mourn its passing, although many of the distillery buildings still stand, including the two pagoda-topped kilns (the third was demolished in 2004, as it was unsafe); the dunnage warehouses, among the oldest in Scotland, are still used for maturing Lagavulin, and other buildings have been converted into small business units.

The distillery began as a malt mill, established before 1824 by Alexander Kerr Mackay with the support of the laird of Islay, Walter Frederick Campbell. By 1825 John Morrison was making whisky here, as a sub tenant, and possibly as a representative of his uncle, Ebenezer Ramsay, Procurator Fiscal of Clackmannan, a relation of the Stein distilling dynasty, with extensive distilling interests around Alloa. The business did not prosper, and in 1833 Uncle Ebenezer sent first his son then his nephew to Islay to investigate. His son reported that the distillery was unworkable; however, the nephew, John Ramsay, reported the contrary.

John Ramsay (1815–92) was 18 years old. After some training as a distiller in Alloa, he was installed as Manager. Within only a few years he was also managing the laird's affairs and became his business partner in some enterprises, including the introduction of the first bi-weekly steamship service between Islay and Glasgow. Ramsay became a leading agriculturalist on the island, Chairman of the Glasgow Chamber of Commerce, Liberal MP for Stirling (1868) and for Falkirk (1874–86). When he died he owned the entire parish of Kildalton, had built numerous farmhouses, estate houses and steadings, and done much to improve the standard of farming on the island.

Port Ellen was continued by his widow and then their son, Captain Iain Ramsay of Kildalton.

SEVENTH 7 RELEASE

NATURAL CASK STRENGTH SINGLE MALT WHISKY

ESTABLISHED. 1825
PORT ELLEN.
ISLAY SINGLE MALT SCOTCH WHISKY

PORT ELLEN DISTILLERY
PORT ELLEN, ISLE OF ISLAY PA42 7AJ

Limited Edition Numbered Bottle No. 0625

One of only 879 bottled in 2007
Distilled in 1979. Aged 28 Years

Matured and Bottled by the Distillers

53.8% vol 70cle

Port Ellen was the first distillery to export whisky direct to the United States.

Keeping the business going after World War I (in which he was wounded) was a struggle, and he was obliged to sell the distillery to his former agent, James Buchanan & Company, in partnership with John Dewar & Sons. When these companies joined D.C.L. in 1925, Port Ellen went with them.

It was put under the management of S.M.D. from 1930 and was immediately closed (for 37 years), although the maltings continued to operate. In 1967 it was rebuilt within the shell of the original distillery, with four mechanically stoked stills (converted to steam heating in 1970) and worm tubs. In 1973 a drum maltings was built adjacent to the distillery, first to supply S.M.D.'s requirement at its three Islay distilleries, and then in terms of a concordat signed by all the Islay and Jura distillers in 1987, to supply at least a portion of their malt requirement so as to keep the maltings in full production.

Port Ellen Distillery was mothballed in 1983, and closed 1987.

Since 2001, Diageo have been bottling small amounts of Port Ellen as 'Special Releases' annually, all from 1978 or 1979. These are highly sought after by collectors.

CURIOSITIES: All the Islay distilleries (except the newcomer Kilchoman) were built on the sea's edge, to facilitate deliveries of barley and coal and the export of whisky. A pier was built at Port Ellen in 1826 and the village grew up around it, named after Eleanor, wife of W.F. Campbell of Islay.

From 1869 the sales side of the business was handled by W.P. Lowrie in Glasgow, the pioneer of cask seasoning with sherry and the original supplier of James Buchanan, whose Black & White brand became so successful he bought Lowrie's in 1906.

Pulteney

REGION
Highland (North)

ADDRESS
Pulteneytown, Wick,
Caithness

PHONE
01955 602371

WEBSITE
www.oldpulteney.com

OWNER
Inver House Distillers

VISITORS
Visitor centre since 2004
in a converted dunnage
warehouse with facility to
fill your own bottle

CAPACITY
1.8m L.P.A.

HISTORICAL NOTES: The awkwardly named
Pulteneytown is the fishing port of Wick. It was
built as a model village between 1800–1820 and was
named after Sir William Pulteney, Director General
of the British Fisheries Society.

Pulteneytown was designed by the leading civil
engineer, Thomas Telford, over the Spey. By the mid
nineteenth century it was the largest herring port in
Europe (and in the world, by weight of catch), used
as a base by over 1,000 boats and attracting 7,000
migrant workers during the season.

The distillery was built in 1826 by James
Henderson, whose family continued to own it until
1920, when it was sold to James Watson & Company
of Dundee, passing to Dewar's three years later, and
thence to the D.C.L., who mothballed it in 1930. In
1951 it was sold to a solicitor, Robert Cumming (who
also bought Balblair), but he soon sold to J. & G.
Stodart, a subsidiary of Hiram Walker & Sons, who
rebuilt it in 1958. It passed to Allied in 1961, and they
sold to Inver House in 1995.

CURIOSITIES: Sir William Pulteney was born in
Dumfriesshire as William Johnstone. He changed
his name when he married the niece of the Earl
of Bath. When she inherited her uncle's fortune,
William became one of the richest men in England.

In his youth in Dumfries he had met a young and
impoverished stonemason, Thomas Telford, and
when he became rich Pulteney became Telford's
main sponsor – under which patronage he became
the leading civil engineer of the day. Indeed Thomas
Telford is often described as 'The Father of Civil
Engineering'. In 1801 Telford devised a master plan
to improve communications in the Highlands, a
massive project that was to last 20 years. It included
the building of the Caledonian Canal along the
Great Glen and redesigning sections of the Crinan
Canal, nearly a thousand miles of new roads, over a
thousand new bridges (including the famous bridge

over the Spey at Craigellachie), numerous harbour improvements and 32 churches.

It is claimed that the unusual flat-topped stills were truncated because they were too tall to fit in the stillhouse (see also *Cragganmore*).

Under the influence of the American evangelist Aimee Semple McPherson in the 1920s, Wick went 'dry' until 1947. To mark the 50th anniversary of repeal, a 12YO Old Pulteney was released, and in 2007 the 60th anniversary was celebrated with a charity ball in aid of the R.N.L.I.

Old Pulteney used to be described as the 'Manzanilla of the North': it was pale and distinctly briny. Today it is positioned as 'the genuine maritime malt', and its bottle and tube are illustrated with trawlers. In 2006 it was the main sponsor of the I.R.C. Scottish (Yacht Racing) Championships. The brand also sponsored Sir Robin Knox-Johnson, who successfully completed his second solo round-the-world race in 2007.

Pulteney is the most northerly distillery on mainland Scotland, and is one of only two named after people (the other being Glen Grant – but see also *Port Ellen*).

RAW MATERIALS: Soft process water and cooling water from the Loch of Hempriggs via the longest lade in Europe: 5.5 miles long, stone built by Telford to supply water to Pulteneytown. Maltings removed 1958; unpeated malt from independent maltsters.

PLANT: Cast iron semi-Lauter tun with a copper canopy (4.94 tonnes). Six cast iron washbacks lined with Corton steel. One wash still, with T-shaped lyne arm and the largest boil-ball in the industry (14,400 litres charge); one boil-pot spirit still – similar in shape to a 'smuggler's kettle' (13,200 litres charge). Both indirect fired by steam coils. Worm tubs.

MATURATION: Mainly bourbon and a small number of ex-sherry casks. All whisky bottled as a single is matured on site in seven dunnage warehouses (30,000 cask capacity).

STYLE: Fruity, oily (almond oil), malty, heavy, meaty – complex.

MATURE CHARACTER: Heavy as new-make, Pulteney becomes lighter and fresher as it matures. The nose is distinctly maritime, with oily notes, and light, fresh fruit. The taste is dry overall, and slightly salty, with traces of nuts. Medium- to light-bodied.

Rosebank Redeveloped

REGION
Lowland

ADDRESS
Camelon Road, Falkirk

OWNER
Diageo plc

CLOSED
1993

HISTORICAL NOTES: Rosebank was established within the maltings of the former Camelon Distillery (known to have been operating 1817–19) by local grocer James Rankine. The situation was ideal: on the bank of the Forth and Clyde Canal a mile from Falkirk, 'placed by the main road, on which there is a constant stream of traffic, and also fronting the canal, where boats and steamers are continually passing to and fro' (Barnard, 1885). Ironically it was the busy main road that caused Rosebank's ultimate demise.

Rankine began his operation in 1840 and enlarged the site five years later. By the 1860s the distillery was managed by his son, R.W. Rankine, and in 1864 he 'entirely rebuilt in modern form' (Barnard), using red brick. The buildings had the canal on one side, the road on the other and were grouped around a courtyard. Next year he demolished the main Camelon Distillery on the other side of the canal and built his maltings there, connecting them to the distillery by a swing bridge. Beyond this was Rosebank House, standing in three acres of gardens, the 'country residence' of Mr Rankine.

Together with Glenkinchie, Saint Magdalene, Grange and Clydesdale Distilleries, Rosebank formed S.M.D. in 1914, under the chairmanship of W.H. Ross, Managing Director of D.C.L., and joined the larger company in 1925.

In spite of its good reputation, the distillery was closed in 1993, largely owing to difficulties of access off the main road.

The main distillery buildings now house a Beefeater restaurant. Some of the warehouses have been demolished and redeveloped, and some have been converted. Part of the distilling equipment is stored on site (one of the stills was vandalised in December 2007 by copper thieves).

In 2013, the 'Camelon Distillery Project', inspired by Gerald Michaluk (founder of the Arran Brewery), received a grant of £500,000 from the Scottish

Government to restore the former Rosebank Distillery site and open The Forth and Clyde Brewery, with visitor facilities and incorporating the Scottish National Brewing and Distilling Centre.

Distilling on the site is forbidden by its former owner, Diageo, until 2017, after which Mr Michaluk hopes to 'restore the distillery to its former glory'.

CURIOSITIES: Rosebank had the highest reputation of any Lowland malt: in the 1890s there was 'an extraordinary demand for the make and many customers had to be content with an allocation of a smaller amount than they had ordered', (Snilly)

The 1920s and 1930s were hard years for the whisky industry. Many distilleries closed. D.C.L. consolidated all their distilleries under the ownership of S.M.D. in 1930, and used the subsidiary to buy up and close down failing distilleries in order to control output and price. By 1935 S.M.D. comprised 51 malt whisky distilleries.

Like many Lowland malt distilleries, Rosebank favoured triple distillation.

Ro

Ro

Roseisle

REGION
Speyside
ADDRESS
Roseisle, Morayshire
PHONE
01343 832100
WEBSITE
No
OWNER
Diageo plc
VISITORS
No
CAPACITY
12.5m L.P.A.

HISTORICAL NOTES: In January 2007 Diageo announced their intention to build a 'super-distillery' near their maltings at Roseisle, to produce a range of whiskies of varying styles for blending purposes.

The distillery was complete by spring 2008, and went into production that autumn. The cost was £40 million. It has the largest capacity of any Scottish malt distillery except Glenfiddich and can produce a range of styles of Speyside spirit – broadly, 'light' and 'heavy' – made possible by being able to switch condensers (from copper to stainless steel) on six of the stills. Roseisle is also a state-of-the-art example of the application of ingenious recycling procedures which together reduce this massive distillery's environmental impact to a fraction of that of a standard distillery.

CURIOSITIES: When the maltings were built at Roseisle in 1979/80 it was planned to build a distillery alongside, but nothing like on this scale.

Examples of Diageo's environmental ingenuity include:

Draff is dried through a belt press mangle, mixed with solids extracted by centrifuge from the pot ale and by-products (culms and rootlets) from the maltings, and used in a bio-mass burner to make steam. About half the required steam comes from this source.

Hot water from the condensers is piped over to the maltings, and those at Burghead, providing hot air to dry the green malt.

Because the malt comes from the adjacent maltings, transport costs, emissions and traffic impact are negligible.

RAW MATERIALS: Malt from Roseisle Maltings. Water from springs on site.

PLANT: Two full-Lauter mash tuns (13 tonnes each). 14 stainless steel closed washbacks (with an extra two to store hot water from the wash still condensers to pre-heat the next wash still charges). 7 plain wash stills (externally heated) and 7 plain spirit stills (heated by steam coils), all with unique parallel-sided pots (wash still charge 20,000 litres; spirit stills 12–15,000 litres). Each of the spirit stills is fitted with interchangeable copper/stainless steel shell-and-tube condensers.

MATURATION: Mainly U.S. oak, from Cambus cooperage. Matured in the central belt.

STYLE: Both light and heavy Speyside.

Royal Brackla

REGION
Highland (East)

ADDRESS
Cawdor, Nairn, Highland

PHONE
01667 402002

WEBSITE
No

OWNER
John Dewar & Sons Ltd
(Bacardi)

VISITORS
By appointment

CAPACITY
4m L.P.A.

HISTORICAL NOTES: A map of Cawdor Estate from 1773 shows a 'malt brewhouse' on the site where Brackla Distillery was built in 1812. The founder was Captain William Fraser of Brackla, an irascible gentleman, who inveighed against smuggling in the district before the Parliamentary Commission of 1821 ('I have not sold 100 gallons for consumption within 120 miles of my residence during the past year, though people drink nothing but whisky') and was repeatedly fined by H.M. Customs & Excise during the 1830s and 1840s for unknown offences!

Notwithstanding this, Fraser was granted a Royal Warrant by William IV in 1835, the first distiller to be so honoured. Andrew Usher & Company, Edinburgh, were appointed agents in about 1844 and became partners in the business. A small amount of Brackla (1/24th part!) went into their blend Usher's O.V.G. after 1860. When the lease was regranted by Lord Cawdor in 1890 to Robert Fraser & Company, the partners were Andrew Usher II and his brother Sir John Usher. They rebuilt the distillery that year, and when Andrew Usher died in 1898, the firm was incorporated as the Brackla Distillery Company Ltd.

John Bisset & Company Ltd of Aberdeen bought the lease in 1926, converted it to a Feu Charter and sold the distillery to S.M.D. in 1943. A major refurbishment took place in 1964/65, when a second pair of stills were installed and all four converted to indirect firing by steam.

Royal Brackla was mothballed from 1985 to 1991, but resumed production that year and underwent a £2 million refurbishment in 1997 – in time for it to be sold to Bacardi, along with John Dewar & Sons the following year.

CURIOSITIES: Cawdor Estate, the land upon which Brackla stands, was made legendary by Shakespeare in *Macbeth*, and the 'Thane of Cawdor' is still the current Earl of Cawdor. The play is set on the rolling

James Augustus Grant, who accompanied Captain Speke on his last journey to discover the source of the Nile in 1860, wrote to his sister that after dinner they felt 'as comfortable as if we had our legs under your table drinking "Brackla"'.

coastal plain between Inverness and Forres (see *Glenburgie*).

Since he could not sell his whisky locally, Captain Fraser advertised in the *Aberdeen Chronicle* of 1828 that he had 'made an arrangement to have this *much admired spirit* sent up [to Aberdeen] by land, when a regular supply can be had weekly'.

Alfred Barnard (1887) mentions that 'the whisky is carted to the station six miles distant by a traction engine which brings back coals from Nairn'.

Brackla has among the longest fermentations of any distillery: an average of 80 hours, with short fermentations at 72 hours and long fermentations at 110 hours.

Little Royal Brackla was bottled by its owners until 2014 when Dewars repackaged the brand and released a number of bottlings at different ages.

RAW MATERIALS: Floor maltings until 1966; unpeated malt from independent maltsters. Process water from Cawdor Burn and cooling water from the Cursac spring.

PLANT: Semi-Lauter stainless steel mash tun (12.5 tonnes per mash). Six Oregon pine and two stainless steel washbacks. Two plain wash stills (20,500 litres capacity, two plain spirit stills (23,000 litres). All indirect fired. All shell-and-tube condensers.

MATURATION: Five bonded warehouses, some dunnage, some racked (built 1975). Today all Dewar's own stock is tankered to Coatbridge and Glasgow for maturation in mainly refill American oak.

STYLE: Green grassy.

MATURE CHARACTER: Fresh, floral and grassy, with notes of cream and light coconut, and sometimes a whiff of smoke. Smooth mouthfeel, with vanilla cream; sweet taste to start, with malt, apples and pears, leafy-floral notes and sometimes spice. Medium-bodied.

Royal Lochnagar

REGION
Highland (East)

ADDRESS
Crathie, Ballater,
Aberdeenshire

PHONE
01339 742700

WEBSITE
www.malts.com

OWNER
Diageo plc

VISITORS
Visitor centre with display
area, a good shop (with
a range of Diageo's rarer
bottlings) and guided
tours

CAPACITY
500,000 L.P.A.

HISTORICAL NOTES: Royal Lochnagar is Diageo's smallest distillery, and its showcase 'Malts Brand Home' used for V.I.P. visits and training, on account of its charm and picturesque location.

Deeside was long a hotbed of illicit distilling. After the 1823 Act, former illicit distillers took out licences and their erstwhile colleagues sometimes turned against them. Such was the case with James Robertson of Crathie, who built a distillery in Glen Feardan, north of the River Dee, and had it burned down. He then built a second distillery called Lochnagar, also on the north bank of the Dee in 1826. It too was burned. He built a third distillery, and again the entire premises were '... reduced to embers. Where or how the fire originated remains a mystery.' (*Aberdeen Journal*, 12 May 1841)

So it was a brave man who built New Lochnagar Distillery on the other side of the Dee in 1845, close to Balmoral Castle, which was leased by Queen Victoria and Prince Albert in 1848, then bought by them and rebuilt. This man was John Begg, whose name was later immortalised by the slogan for his blended whisky, 'Take a peg of John Begg'.

Knowing how interested Prince Albert was in 'things mechanical', Begg sent an invitation to Balmoral to come and inspect his 'works' two days after the royal party arrived at Balmoral for the first time. The Prince arrived next day, accompanied by the Queen and their three eldest children; they toured the distillery and he gave them all a dram. It must have been good stuff, for it was followed within days by the award of a Royal Warrant. Before long the distillery was calling itself Royal Lochnagar.

The distillery passed to John Begg's son and grandson, and then a family trust and, after 1902, a private limited company.

Lochnagar is a mountain that rises up behind the distillery and the Balmoral Estate. It was immortalised by Lord Byron: 'Oh, for the crags that are wild and majestic The step frowning glories of dark Loch-na-gar.'

During World War I the Directors approached D.C.L., which acquired the family's shares in 1916.

CURIOSITIES: The distillery takes its name from the mountain which rises close by, and the loch close to the summit of the mountain.

Royal Lochnagar was Lord Macfarlane of Bearsden's favourite whisky. As Sir Norman Macfarlane, Chairman of U.D., he had initiated the first proprietary bottlings of the malt.

RAW MATERIALS: Process water from springs in the foothills of Lochnagar; cooling water from two reservoirs. Unpeated malt with POtaiele maltings.

PLANT: Open-topped cast iron rake and-plough mash tun (5.4 tonnes per mash). Three Scottish larch washbacks (two of 37,000 litres; one of 18,000 litres). One plain wash still (6,300 litres charge), one plain spirit still (4,200 litres charge), both indirectly heated by steam. Worm tubs.

MATURATION: American and European oak puncheons and butts. One dunnage warehouse on site, holding 31,000 casks (formerly the distillery's maltings). The remainder stored at Glenlossie Distillery.

STYLE: Light and grassy. Small stills and worm tubs should produce a heavy meaty malt, but the way they are operated (with a low-volume charge to the spirit still, air rests to cool the copper, and warm water in the worm tub) achieves this specification.

MATURE CHARACTER: Light toffee and planed hardwood on the nose, with some piney notes, boat varnish and linseed oil; dry overall. Sweet to start, then acidic, with an attractive lingering sandalwood aftertaste. Medium-bodied.

Scapa

REGION
Highland (Island)

ADDRESS
St Ola, Kirkwall, Orkney

PHONE
01856 876585

WEBSITE
www.scapamalt.com

OWNER
Chivas Brothers

VISITORS
Visitor centre and shop
since April 2015

CAPACITY
1.1m L.P.A.

HISTORICAL NOTES: The distillery stands beside the Lingro Burn and overlooks the spacious anchorage of Scapa Flow, where the German High Seas Fleet scuttled itself in 1919, and where a daring German submarine sank the battleship *Royal Oak* in 1939.

A local minister reported in 1701 that the parish held a large and ancient drinking cup, which had supposedly belonged to St Magnus (eleventh century) and 'full of some strong drink' was presented to Bishops of Orkney upon their arrival. 'If the Bishop drank it out, they highly praised him, and made themselves believe that they should have many good and fruitful years in his time.'

Scapa Distillery was built by John Townsend, a Glasgow blender, in 1885 and operated until 1919 when it was narrowly saved from total destruction by fire, thanks to sailors from the Grand Fleet, anchored in Scapa Flow, forming a chain of buckets from the sea. It then passed through various hands, until it was bought by Hiram Walker in 1954; they rebuilt it five years later, installing a Lomond-style wash still (see *Inverleven, Miltonduff*). It was again modernised in 1978, when the plates within the still were removed, so it now operates like a straight-sided traditional still. Again the distillery was mothballed in 1994, and from 1997 production was sporadic and done by staff from Highland Park Distillery nearby.

Ownership passed to Allied Distillers when they bought Hiram Walker in 1986/87, and in 2004 they surprised the industry by undertaking major refurbishment (at a cost of £2.1 million), launching the first proprietary bottling and establishing a fan club. Production was again halted in 2005 to allow completion of the refurbishment programme; later that year ownership passed to Pernod Ricard/Chivas Brothers, who resumed production in October 2005, 120 years after it opened.

CURIOSITIES: When Alfred Barnard visited Scapa in 1886 he found it to be 'one of the most complete little Distilleries in the Kingdom'.

The large water wheel, driven by the Lingro Burn, supplied power to the distillery.

Scapa has operated a Lomond wash still from 1959 to the present day. It is now the only distillery to have such a still (although adjusted). It is unusual among the island distilleries in specifying unpeated malt.

RAW MATERIALS: Unpeated malt from Kilgours, Kirkcaldy; own floor maltings removed 1962. Cooling water from the Lingro Burn; process water from springs and the Coltland Burn.

PLANT: Stainless steel mash tun (3.76 tonnes per mash); six stainless steel washbacks. One remade Lomond style wash still, now with no internal plates and a purifier (13,500 litres charge); one plain spirit still (9,000 litres charge). All indirect fired. Shell-and-tube condensers.

MATURATION: Mainly refill U.S. hogsheads, some first-fill.

STYLE: Heather pollen, honey and light spice.

MATURE CHARACTER: Scapa has a maritime character, with faint saltiness on the nose, together with floral/grassy notes and some scented wood. The taste is dryish and lightly spicy, with some vanilla and toffee notes. Medium-bodied.

Speyburn

REGION
Speyside

ADDRESS
Rothes, Moray

PHONE
01340 831213

WEBSITE
www.speyburn.com

OWNER
Inver House Distillers

VISITORS
By appointment

CAPACITY
4.2m L.P.A.

HISTORICAL NOTES: Speyburn is a picturesque distillery tucked snugly into a steep wooded glen just outside Rothes. It was designed for John Hopkins & Company (owners of Tobermory Distillery) by the leading distillery architect Charles Doig of Elgin, and built from stones quarried from the adjacent riverbed. It opened in 1897. Since the site was narrow, Doig also installed a 'Hennings pneumatic maltings' – the first drum maltings in any malt distillery – since it took up less ground-space. This operated until 1968. The wooden malt hoppers are still in use.

John Hopkins & Company joined D.C.L. in 1916 and Speyburn was licensed to the now forgotten Leith blenders John Robertson & Sons. It was operated by S.M.D. from 1962; they converted the stills to indirect firing that year and rebuilt and re-equipped the mashhouse in 1974 (although not the stillhouse, which remains as Doig designed it). Inver House Distillers bought Speyburn in 1991.

Speyburn is Inver House's best-selling malt, with over 500,000 bottles sold in 2014, especially popular in the U.S.A. In 2014/15 the distillery was expanded, with a new mash tun and 15 new washbacks. The existing wash still was converted into a spirit still, and a new, larger wash still installed.

CURIOSITIES: Speyburn opened on the last day of 1897, the year of Queen Victoria's Golden Jubilee. The men worked all night in a violent snow storm and in a stillhouse that had no windows. Only one cask bore the historic date.

Speyburn is little known in most markets, but it is in the top six best-selling malts in the U.S.A., and number one in Finland!

RAW MATERIALS: Own drum maltings until the 1967; now unpeated malt from independent maltsters. Soft process water from the Granty Burn, a tributary of the River Spey; cooling water from the Broad Burn.

PLANT: Full-Lauter mash tun (six tonnes). Four Douglas fir washbacks, 15 stainless steel washbacks. One plain wash still (15,000 litres charge), two spirit stills (one 12,500 litres charge, the other 11,500 litres), all indirect fired since 1962. The spirit stills connected to a worm tub (the pipe is over 100 metres long).

MATURATION: Mainly ex-bourbon casks, some European. 80% of spirit tankered and filled into cask at Airdrie; casks for bottling as a single matured on site in a two-floored warehouse, racked three high.

STYLE: Speyside: estery, floral, grassy, citric, but with a meaty note.

MATURE CHARACTER: A lightweight Speyside style: fresh and floral/fruity on the nose, with cereal notes. Predominantly sweet to taste, with green apples, pear drops and herbal-heathery notes. Light-bodied.

Speyside

REGION
Speyside
ADDRESS
Tromie Mills, Glentromie,
Kingussie, Highland
PHONE
01540 661060
WEBSITE
www.speysidedistillery.
co.uk
OWNER
Speyside Distillery
Company Ltd (Harvey's of
Edinburgh)
VISITORS
By appointment
CAPACITY
600,000 L.P.A.

HISTORICAL NOTES: After being demobilised from the Royal Navy (submarines), George Christie joined a former naval colleague as junior partner in the Glasgow firm of whisky brokers W.R. Paterson & Company. Three years later (1949), they bought the whisky broker Alexander McGavin & Company (Glasgow) Ltd at auction from the Inland Revenue, in partnership with one Sandy Grant. George was appointed Managing Director, and in 1955 and 1964 bought out the other shareholders.

In 1955 he also founded the Speyside Distillery & Bonding Company, and next year bought Old Milton Estate, Glen Tromie, three miles from Kingussie. In particular he was interested in Tromie Mills, a barley mill dating back to the eighteenth century, which had been run for generations by a local family, and continued until 1965. When restoration of the building commenced in 1967, the mill and water wheel were retained, and are still in working order. The first spirit ran on 12 December 1990, while the formal opening of the distillery was celebrated with a lunch on 20 September 1991.

George Christie also founded Strathmore Distillery at Cambus, Alloa, in 1957 (see *North of Scotland* entry).

The company's HQ is in Rutherglen, Glasgow, where most of the make is matured, blended and bottled. George Christie sold his shares in Speyside Distilleries in 2000 to his son, Ricky, Ian Jerman and Sir James Ackroyd, who became Chairman of the company. He died in 2011, and in September 2012 the distillery was sold to Harvey's of Edinburgh Ltd, a company owned by John Harvey McDonough, and which has close connections with the Taiwanese company Vedan (the world's largest producer of monosodium glutamate). Harvey's have been selling malt in Taiwan since the 1990s, including the Spey brand, which by 2011 was the third most popular malt in this important market. The range

The rebuilding of Speyside Distillery was the work of one man, the local dry-stane dyker Alex Fairlie. It took him 20 years.

of expressions from Speyside was substantially increased during 2014/15.

CURIOSITIES: Closer to the source of the Spey than any other distillery, on the bank of the Spey tributary, the River Tromie, and close to the village of Drumguish (pronounced 'Drumooish').

There was an earlier 'Speyside Distillery' in Kingussie. Founded in 1895, it ceased production in 1905 and was demolished in 1911. The present distillery appeared in the TV series *Monarch of the Glen* as Lagganmore Distillery.

RAW MATERIALS: Soft process and cooling water from the River Tromie, via the former mill lade. Traditionally unpeated malt from independent maltsters, with small annual matches of heavily peated malt.

PLANT: Stainless steel 'Glen Spey' semi-Lauter tun with a stainless steel dome (four tonnes per mash, the last one built by Newmill Engineering, its inventor). Four stainless steel washbacks. One plain wash still (10,000 litres charge), one plain spirit still (6,000 litres charge). Indirect fired by steam kettles. Shell-and-tube condensers.

MATURATION: Mix of American barrels and remade hogsheads, with a small number of sherry casks. No bonded warehouses on site; matured in Rutherglen, Glasgow, where the bottling plant is located.

STYLE: Light, soft, sweet, vanilla, hint of liquorice.

MATURE CHARACTER: A light style of Speyside: predominantly cereal-like, with fresh apples and pears, and faint floral notes. The taste is sweet, with cereal and fruity notes. Light-bodied.

Springbank

REGION
Campbeltown

ADDRESS
Longrow, Campbeltown,
Argyll

PHONE
01586 551710

WEBSITE
www.springbankdistillers.
com

OWNER
Springbank Distillers
(J. & A. Mitchell)

VISITORS
By arrangement

CAPACITY
750,000 L.P.A.

HISTORICAL NOTES: Springbank is the only Scottish distillery established in the nineteenth century which is still in the ownership of its founding family.

In all likelihood it was originally called Longrow Street Distillery (too easily confused with Longrow Distillery next door, it soon became Springbank), and is said to have been founded in 1828 by one William Reid. He was related by marriage to the Mitchells, a well-known local family who had arrived in Kintyre from the Lowlands around 1600. Indeed, Archibald Mitchell may well have been distilling illegally on the site before Reid built his distillery; what is certain is that he soon ran into financial difficulties and ownership passed to his in-laws, John and William Mitchell, in 1837 (their brothers, Hugh and Archibald, had founded Riechlachan Distillery nearby, which operated from 1825 to 1934). By the following year they were selling whisky to John Walker of Kilmarnock at 8/8d (44p) per gallon.

The brothers quarrelled and in 1872 William left to join his other brothers at Riechlachan. John brought in his son Alexander as partner, the firm becoming J. & A. Mitchell, as it is today.

Springbank closed from 1926 to 1933, during the Great Depression, but was one of only three Campbeltown distilleries to survive the 1920s, when 17 distilleries closed in the town. The other survivors were Glen Scotia and Riechlachan, the latter closing in 1934. The present Managing Director, Mr Hedley Wright, is John Mitchell's great-great-grandson.

In 2004, Hedley Wright opened Glengyle Distillery, down the road from Springbank, which had been founded originally in 1872 by his great-great-uncle (see entry).

It is the most traditional of all Scottish malt distilleries and makes three styles of malt spirit: Springbank, Longrow and Hazelburn, each with differently peating levels (see below).

CURIOSITIES: As well as distilling, John Mitchell was 'a rare judge of sheep and Highland cattle . . . tenant at one time of no fewer than seven farms', according to his obituary. He died in 1892, aged 91.

The *Daily Mail* of 16 June 1974 reported that an hotelier in Galashiels was displaying a bottle of 50YO Springbank to customers at 10p a viewing: 'He paid £29 for it – the wholesale value of the whisky . . . It's a sobering thought when he tells you what it would cost for a dram – if the liquid gold was up for sale (which it isn't). Without so much as a blush, he puts a price tag of £2 on a fifth (of a gill) measure.' How times have changed; a 50YO Springbank (probably from the same batch of 36 bottles) fetched £3,800 at auction in 2001!

Springbank malts all its barley (a proportion of which is grown locally) in its own floor maltings, using locally cut peat. Its mash tun is rake-and plough; it has three stills, and employs a complex near-triple distillation; its wash still is direct fired by oil, and fitted with rummagers, as well as being indirect fired by steam. There is a worm tub on one of the spirit stills. It bottles on site. The whole place has an antique air, with not a computer to be seen.

J. & A. Mitchell bought the well-established independent bottler Cadenhead's (originally of Aberdeen) in 1969. Around the same time, the site of Longrow Distillery (1824–96), adjacent to Springbank, was acquired. One of its warehouses is still used, and another houses Springbank's bottling plant; the site of the original stillhouse is now a car park. In 1973, the first bottling of Longrow was made from heavily peated malt in Springbank's own stills, and this was joined in 1997 by the first release of Hazelburn, triple distilled from unpeated malt. Hazelburn itself was another Campbeltown distillery, which closed in 1925.

Frank McHardy, who was manager at Springbank from 1977 to 1986 and 1996 to 2013, is one of the 'senior' figures in the whisky industry. He started his career at Invergordon in 1963, worked in several distilleries, then as Master Distiller at Bushmills in Northern Ireland from 1986 to 1996.

RAW MATERIALS: Soft water from Crosshills Loch. Own floor maltings; Springbank malt dried for six hours over peat (cut from near Campbeltown Airport) and then for 18–24 hours over hot air. Longrow malt is dried over peat for around 27 hours, then hot air. Hazelburn malt has no peating.

PLANT: Cast iron open-topped traditional mash tun (3.64 tonnes per mash). Six Scandinavian boatskin larch washbacks. One plain wash still (11,000 litres charge) direct fired by oil burners and indirect fired by internal steam coils (unique in Scotland). Rummager. Two plain spirit stills, indirect fired (7,500–8,000 litres charge). Worm tub on first spirit still; condensers on wash and second spirit stills.

Longrow malt is dried entirely over peat from Pitsligo (Aberdeenshire) for around 27 hours; Hazelburn malt is unpeated; Springbank 2.5 times distilled. 80% of the low wines from the wash still charges the first spirit still in the usual way, and the distillate goes to the feints receiver, where it mixes with the foreshots and feints from the third still. 20% goes straight to the low wines and feints charger of the third still, where it is mixed with the distillate from the feints receiver. This is then distilled in the third still, with feints and foreshots going to the feints receiver.

MATURATION: Mainly first-fill ex-bourbon barrels, followed by first-fill ex-sherry; around 20% refill casks; also Madeira, port and Demerara rum casks. Six bonded warehouses on site; four traditional dunnage, two racked seven-high. The entire make is matured on site.

STYLE: Springbank: light peatiness, oily, sweet and heavy. Longrow: heavy, sweet and distinctly peaty. Hazelburn: light, sweet, with hay and malt notes.

MATURE CHARACTER: Springbank needs long aging, and takes it superlatively well. At 10–15yo it only has 'potential': hints of strawberries and cherries and bananas on the nose, with distinct smokiness. A creamy mouthfeel, with some butterscotch and mint, some sweet malt, a light smokiness. Medium-bodied, becoming fuller with age.

St

Starlaw Grain Distillery

ADDRESS
Starlaw Road, Bathgate,
West Lothian

PHONE
01506 468550

WEBSITE
No

OWNER
Glen Turner Company
(La Martiniquaise)

VISITORS
No

CAPACITY
25m L.P.A.

HISTORICAL NOTES: Following the huge success of their blended Scotch whiskies, Label 5 and Sir Edward's in France (mainly), and of Glen Turner single malt from Glen Moray Distillery, which the company bought in 2008. La Martiniquaise opened this substantial grain whisky distillery without any fanfare in 2011, complete with a blending and bottling plant. It is the first grain whisky distillery to be built on a greenfield site since 1964, and was designed by the American company the Colorado Group, which was awarded the Scottish Construction Centre's 'exceptional performance award for successful design and build' in 2010.

RAW MATERIALS: The cereal base to date has been predominantly home-grown wheat and malted barley. The distillery is currently developing the capability to rapidly change between maize and wheat as a main cereal crop, to optimise efficiency.

PLANT: A very modern design philosophy has delivered an exceptionally efficient end-to-end process that allows the site to minimise its carbon footprint and meet the organisation's objectives in terms of sustainable development. The main distillation columns were designed and installed by Frilli of Italy and produce a consistently clean spirit. The facility was designed with additional distillation capacity for the production of neutral spirit.

MATURATION: Maturation takes place exclusively in ex-bourbon barrels across the site's 21 maturation warehouses. Each has the capacity to hold 23,500 casks.

St Magdalene Redeveloped

REGION
Lowland
ADDRESS
Linlithgow, West Lothian
LAST OWNER
D.C.L./S.M.D.
CLOSED
1983

HISTORICAL NOTES: The first recorded licensee was Adam Dawson in 1797. He was a spokesman for the Lowland distillers and was succeeded by A. & J. Dawson in 1829.

The distillery stood beside the main road between Glasgow and Edinburgh, and its situation was greatly enhanced by the opening of the Union Canal between the two cities in 1822, and the arrival of the railway in 1842. It was a sizeable enterprise, with four stills (converted to indirect firing in 1971) and 19 warehouses, including one built of brick of 'enormous proportions' (Barnard).

A. & J. Dawson was incorporated in 1891 but was in difficulties by 1912, when a liquidator was appointed. A new company was formed later the same year with the same name, under the owner-ship of D.C.L., John Walker & Sons and J.A. Ramage Dawson.

In 1914 the Directors formed S.M.D. with four other Lowland distilleries (see *Rosebank*).

Saint Magdalene was one of the many casualties of world recession of 1983, when S.M.D. was obliged to close distilleries 'to bring the level of maturing stock into line with anticipated level of future sales'. The site has been redeveloped as residential flats.

CURIOSITIES: The name derives from the land near Linlithgow known as Saint Magdalene's Cross, upon which the distillery was built in the late eighteenth century. It was once the site of a hospital of the same name and of an annual fair. The distillery was also known as Linlithgow.

Linlithgow was famous for its water. An old rhyme runs:

Linlithgow for wells,
Glasgow for bells,
Peebles for clashes and lees (i.e. 'lies')
And Falkirk for beans and peas

In the sixteenth century it was a centre of milling and malting, and in the eighteenth century for brewing and distilling. Saint Magdalene's water for cooling and for driving the water wheel came from the Union Canal; process water came from the town's supply, which originated in Loch Lomond.

The gigantic shell of Linlithgow Palace towers above Linlithgow Loch. A royal residence from the twelfth century, the existing ruin dates from 1425 to 1630. Mary, Queen of Scots was born here in 1542.

Colonel Ramage Dawson, who died in 1892, had 'extensive and valuable coffee plantations in Ceylon', an estate in Kinross-shire and the colonelcy of the Haddington Artillery, as well as owning the distillery.

Strathclyde Grain Distillery

ADDRESS
Moffat Street, Glasgow
PHONE
01389 724205
OWNER
Chivas Brothers
CAPACITY
39m L.P.A.

1927 was not the most auspicious year to open a distillery, a time of industrial depression in the U.K. and Prohibition in the U.S.A.; nevertheless, Seager Evans had confidence in the future.

HISTORICAL NOTES: Strathclyde Grain Spirit Distillery was built in 1927 by Seager Evans, the long established (1805) firm of London gin distillers, in order to secure supplies of spirit for rectification. It stands on the south bank of the River Clyde in Glasgow, and draws its water from Loch Katrine in the Trossachs.

The move towards Scotch whisky production was boosted in 1936 by the acquisition of the wine and spirits merchant W.H. Chaplin & Company, which had bought the well-known brand name Long John from the successors to Long John Macdonald of Ben Nevis Distillery in 1911. Next year Seager Evans bought Glenugie Distillery in Peterhead.

In 1956 Seager Evans was bought by the American distillers Schenley Industries Inc. of New York State, and this provided a welcome injection of capital at a time when blended Scotch was taking off world-wide. Ownership of the Scottish distilleries was transferred to Strathclyde & Long John Distilleries Ltd, which soon became simply Long John Distillers, and after 1970 Long John International. In 1975, this was sold to Whitbread & Company Ltd., the brewers, whose spirits interests were bought by Allied-Lyons plc in January 1990 for £454 million.

Strathclyde was rebuilt between 1973 and 1978, with two columns for grain whisky production and five for neutral spirit. Having closed their Dumbarton Distillery in 2002, Allied spent more than £7 million increasing Strathclyde's capacity from 32 to 39 million litres per annum. When Allied was broken up in 2005, Strathclyde Distillery went to Chivas Brothers.

CURIOSITIES: In 1956/57 a new malt distillery, named Kinclaith, was built within Strathclyde Distillery to provide fillings for Long John. After Long John International was sold to Whitbread, Kinclaith was dismantled in 1976/77 to make way for an enlarged Strathclyde (See entry p.257).

RAW MATERIALS: Wheat from independent merchants.

PLANT: Two column stills for grain whisky; five for neutral spirit.

MATURATION: Filled into butts, hogsheads and barrels, and matured on site.

STYLE: Light and sweet, slightly creamy with bubblegum notes.

Strathearn

ADDRESS
Bachilton Farm Steading,
Methven, Perth

PHONE
01738 840100

WEBSITE
www.strathearndistillery.
com

OWNER
Tony Reeman-Clark,
David Lang and David
Wight

VISITORS
Visitor centre and
restaurant

CAPACITY
11\,11V8 L.i .A.

The draff residues go to a local farm which raises wild boar.

HISTORICAL NOTES: Notwithstanding Abhainn Dearg (where the stills are substantially larger), Strathearn is Scotland's first micro-distillery. It commenced production in October 2013 and is the brain-child of Tony Reeman-Clark, David Lang and David Wight, following a conversation at Whisky Fringe in Edinburgh. Like Daftmill, the distillery has been built within an attractive 160-year-old stone farm steading; the stills, which are of the alembic type, were made in Portugal by Hoga, originally designed to make the fortified spirit for sherry industry. Such stills are now widely used for whisk(e)y production in Europe and North America.

The distillery offers a limited number of 50-litre casks for sale, and also distills its own gins, Heather Rose, Classic and Oaked Highland (the latter designed to be drunk straight), from a base of neutral grain spirit, rectified using a separate lyne arm on the spirit still fitted with a botanical basket.

CURIOSITIES: Strathearn probably has the longest fermentation time of any Scottish distillery: 96 hours.

The company founders also founded the Scottish Craft Distilling Association.

RAW MATERIALS: Soft water from Loch Turret. Currently, malt from an independent maltster, but there are plans to install a floor maltings and meet the requirement with barley grown on the farm.
The majority will be unpeated, but a little peated at around 20ppm phenols.

PLANT: Stainless steel mash tun (half tonne, charged 425kg). Two stainless steel washbacks, with cooling jackets. One alembic-style wash still (800–850 litres charge); one alembic-style spirit still (425 litres charge). Shell-and-tube condensers.

MATURATION: Octaves (50 litre) made from virgin French and American oak and ex-sherry casks. Port, Merlot and others also planned. Currently 50% virgin oak and 50% first-fill ex-sherry octaves.

STYLE: Light and fruity spirit; shortbread, 'with a hint of dark chocolate sprinkled with dried orange peel'.

Strathisla

REGION
Speyside

ADDRESS
Keith, Moray

PHONE
01542 783044

WEBSITE
www.maltwhisky
distilleries.com

OWNER
Chivas Brothers

VISITORS
Visitor centre

CAPACITY
2.45m L.P.A.

HISTORICAL NOTES: The distillery was founded in 1786 by a local worthy, George Taylor (flax-dresser, postmaster, banker), in partnership with the Duke of Gordon's factor, with a single 40 gallon still. Originally the distillery was named Milltown, which became Milton after 1825 (although the make was called 'Strathisla'), and in 1830 it came into the hands of William Longmore, banker and grain merchant. In due course it passed to his son-in-law, J. Geddes Brown, who incorporated the business as William Longmore & Company Ltd. Most of the shares were subscribed by local people.

It is interesting to note that Sir Robert Burnett & Company of London (the gin makers) were bottling 'Longmore's Strathisla' as a single in the 1880s, in 'immense quantities' (*Moray & Nairn Express*, 1885): 'Wherever one travels one finds that those who know what good whisky is speak in the most glowing and appreciative terms of the produce of Milton Distillery at Keith.'

Longmore & Company owned Milton until 1950, when it was sequestrated and bought for £71,000 by James Barclay, who immediately sold to the Seagram Company. Thenceforward the distillery was named 'Strathisla', after its product.

At the time Longmore & Company was wound up, Milton was controlled by a shady London financier named Jay Pommeroy, who had siphoned off large stocks of mature whisky and 'disposed of them in a manner which was calculated not to attract any taxation liability', according to the judgement in the Court of Session: that is, he sold them on the black market for huge profit, there being a chronic shortage of mature whisky at the time.

Seagram's immediately expanded production and installed two new stem-heated stills alongside the coal-fired pair in 1965, as well as building extensive warehousing nearby. Seagram's whisky interests were acquired by Pernod Ricard in 2001,

Like many others, Strathisla had a distillery cat to keep rodents under control. A recent one arrived in 1993 as a stowaway in a cargo of casks from Kentucky, starving and drunk from the fumes of the barrels. She was accordingly named 'Dizzy'!

and Strathisla is managed by their subsidiary Chivas Brothers.

CURIOSITIES: In January 1876, the *Banffshire Journal* reported a fire at Milton Distillery, which consumed the byre next to the threshing mill: 'out of 66 cows in the byre, 30 perished in the fire, at a loss of £700; along with 500 quarters of barley, an excellent threshing machine and a steam engine ... The damage was estimated at £3,800.' Three years later there was an explosion in the malt mill, a common peril caused by 'a small piece of stone coming in contact with the cylinder [of the mill] and originating a spark which set fire to the fine, powdery and very explosive [dust from the mill]'.

A conspicuous feature of the distillery is the water wheel, installed in 1881 to provide power, and used until 1965 to drive the rummagers in the wash stills.

Milton/Strathisla has the highest pagoda roofs of any Scottish distillery (although they do not look like it, since the distillery is small) and, unusually, they were not designed by Charles Doig (the inventor of the pagoda roof and doyen of distillery building) but by another distillery architect, John Alcock.

James Barclay (1886–1963) was one of the most colourful characters in the whisky industry during the post-war decades. He started as a clerk at Benrinnes Distillery in 1902, then worked for Peter Mackie at White Horse. With a partner, he bought George Ballantine & Son in 1919, and J. & G. Stodart (blenders) in 1922, and immediately started to arrange distribution in the U.S.A. – which at the time was 'dry'. In Canada, he became friendly with Harry Hatch, the owner of Hiram Walker, Gooderham & Worts, the largest distiller in Canada, and sold Ballantine's to him in 1935. Next year he bought Miltonduff and Glenburgie Distilleries on behalf of Hiram Walker, and joined the Board

Strathisla is the oldest working distillery in the Highlands, and is one of the most attractive in Scotland, tastefully restored in 1995 and a listed building.

of Hiram Walker (Scotland) Ltd. Somewhat mysteriously, he resigned in 1937 and even more mysteriously transferred his allegiance to Walker's main rival, Seagram, paving the way for their entry to the Scotch whisky industry by the purchase by Chivas Brothers, of Aberdeen, and then of Milton/Strathisla.

RAW MATERIALS: Floor maltings until 1961; now unpeated malt from Paul's, Buckie. Production water from Fons Bulliens Well, which, it is believed, was used by monks for brewing beer in the thirteenth century.

PLANT: Stainless steel, traditional rake-and-plough mash tun, with a copper dome (5.12 tonnes). Ten Oregon pine washbacks. Two boil-pot wash stills (12,500 litres charge); two boil-ball spirit stills (7,000 litres charge). Direct fired until 1990s, now all indirect fired. Shell-and-tube condensers.

MATURATION: Mainly refill U.S. hogsheads, some first-fill.

STYLE: Sweet, fruity and estery.

MATURE CHARACTER: Strathisla is a rich malt at its best, with deep fruity scents, including apricots and plums, sweet malt, sandalwood and even a whiff of smoke. It has a pleasant, full texture in the mouth and a sweet, sherried taste, becoming tannic/dry towards the end. Medium-long finish. Medium-bodied.

Strathmill

REGION
Speyside
ADDRESS
Keith, Moray
PHONE
01542 883000
WEBSITE
www.malts.com
OWNER
Diageo plc
VISITORS
No
CAPACITY
2.6m L.P.A.

HISTORICAL NOTES: Strathmill began life in 1891 as Glenisla-Glenlivet, within a converted flour mill, itself built in 1823 and called Strathisla Mills. Its name was changed four years later when it was bought by the London gin distillers W. & A. Gilbey (who already owned Glenspey Distillery and would later buy Knockando Distillery).

In 1962 Gilbey's merged with United Wine Traders (including Justerini & Brooks) to form International Distillers & Vintners (I.D.V.); six years later Strathmill was expanded to four stills, and purifiers were added to increase reflux and lighten the spirit. I.D.V. was bought by the brewer Watney Mann in 1972, and both were immediately absorbed by the Grand Metropolitan Hotel Group. Grand Metropolitan and Guinness merged to form Diageo in 1997.

CURIOSITIES: It was from here that the first whisky tanker was used to carry spirit to blenders. The tanker was named *Whisky Galore*, which probably dates it to the 1950s.

'[Until 1905] Gilbey's refused to admit that anything but pure malt could be called "Scotch whisky", but it was gradually appreciated that malt whisky was too heavy for a southern climate.' (*Merchants of Wine*, Alec Waugh)

RAW MATERIALS: Process water drawn from a spring on site; cooling water from the Isla Burn. Non-peated malt from Burghead in Elgin.

PLANT: Stainless steel Lauter tun with a peaked canopy (nine tonnes per mash); six stainless steel washbacks. Two boil-ball wash stills (11,000 litres charge) and two boil-ball spirit stills (6,700 litres) with purifiers. Shell-and-tube condensers.

MATURATION: Refill sherry butts and refill hogsheads; six warehouses on site.

STYLE: Light, grassy, malty.

MATURE CHARACTER: A light Speyside style, designed for blending (principally the J. & B. blends). The nose is sweet and estery, with floral and fresh-fruity notes (reminiscent of tangerine, oranges and sweet apples). The taste is also sweet, with fruity notes. Medium- to light-bodied.

Talisker

REGION
Island (Skye)

ADDRESS
Carbost, Isle of Skye

PHONE
01478 614308

WEBSITE
www.malts.com

OWNER
Diageo plc

VISITORS
Visitor centre and shop

CAPACITY
2.7m L.P.A.

HISTORICAL NOTES: Founded in 1830 by Hugh and Kenneth MacAskill, tacksmen (substantial tenant farmers), who had acquired the lease of Talisker House (where Johnson and Boswell stayed in 1764) and its associated estate from Macleod of Macleod.

Importing barley and coal and exporting whisky, all by sea, was costly. The MacAskills sold to the North of Scotland Bank in 1848, and two subsequent licensees went bankrupt before the distillery was bought by Roderick Kemp in 1879 (see *Macallan*). With his partner, Alexander Grigor Allan (see *Dailuaine*), the distillery was extended in 1900 and the pier and cottages built. It came under the control of the Big Three and D.C.L. in 1916, and was absorbed by the latter in 1925.

Until the stillhouse was rebuilt in 1961/62, following a devastating fire, Talisker was triple-distilled. Today's distilling regime achieves the same goal by increasing reflux in the spirit stills (see below). Floor maltings closed in 1972. The visitor centre opened in 1988.

CURIOSITIES: On its way to the worm tub, the spirit stills' lye pipes describe an inverted 'U', just before which is a purifier (a return pipe to the still). This greatly increases the amount of reflux – a past manager estimated that 90% of the vapour that reached the U-bend was returned and redistilled.

A unique characteristic of Talisker is the peppery, even chilli-peppery 'catch' as you swallow. Nobody knows where it comes from.

Talisker is one of only a handful of malts bottled as singles in the early twentieth century. Until it was selected as a Classic Malt in 1988 it was bottled at 8YO, thereafter at 10YO. An oddity of the make is that it is traditionally bottled at slightly higher than standard strength. Very few independent bottlings have been done, since the malt is extensively used in the Johnnie Walker blends. In recent years Diageo have promoted Talisker strongly and have

*Robert Louis
Stevenson
famously remarked
in 1887:
'The King o' drinks
as I conceive it –
Talisker, Islay or
Glenlivet.'*

introduced a number of new expressions. As a result, the malt stands at number ten in the list of global best-sellers.

RAW MATERIALS: Medium-peated malt (18–25ppm phenols) from Glen Ord Maltings (7–8 in final spirit). Process water from a burn rising on Cnoc nan Speireag (Hawkhill) behind the distillery; cooling water from Carbost Burn.

PLANT: Semi-Lauter mash tun (eight tonnes); six wooden washbacks. Two boll-ball wash stills (14,000 litres charge), with a horseshoe-shaped U-bend on the lye pipe before it enters the worms tub, and a purifier pipe, leading to greater reflux; three plain spirit stills (11,300 litres charge). Indirect fired by steam coils and pans. Worm tubs.

MATURATION: Mainly refill U.S. hogsheads; a small amount of refill European wood. 4,500 casks on site; remainder filled and matured in the Central Belt

STYLE: Smoky and spicy, with maritime notes and high pungency. Its keynote flavour (for me) is chilli-pepper in the back of the throat.

MATURE CHARACTER: Talisker is always bottled at slightly higher strength than other whiskies, which enhances its pungency. It is elemental and maritime. Beaches, seaweed, salt spray, with spice, dried fruits and distant bonfires. The taste is sweeter than expected, with rich fruits, fragrant smoke and chilli pepper. Full-bodied.

LIMITED EDITION
0235/0493

TALISKER
THE ONLY SINGLE MALT SCOTCH WHISKY
FROM THE ISLE OF SKYE

NATURAL CASK STRENGTH
30 YEARS OLD
SINGLE MALT SCOTCH WHISKY

Tamdhu

REGION
Speyside

ADDRESS
Knockando, Aberlour,
Moray

PHONE
01340 870221/ 872200

WEBSITE
www.tamdhu.com

OWNER
Ian Macleod
Distilleries Ltd

VISITORS
No

CAPACITY
4m L.P.A.

In the early 1980s, labels of Tamdhu 10YO bore the legend: 'Now calmed in deep pools to reflect softly on the day; soon to well the tumbling torrents of the Spey'.

HISTORICAL NOTES: In 1896 three distilleries were planned for Knockando parish: Tamdhu, Knockando and Imperial. The moving force behind Tamdhu was William Grant, a Director of Highland Distilleries and agent for the Caledonian Bank in Elgin. The place he chose beside the Knockando Burn was well-supplied with pure water from springs, and according to local tradition had been used by illicit distillers in days gone by. As important, it was adjacent to the Strathspey railway. He soon raised the funds required to build the distillery (£19,200) from fifteen whisky brokers and blenders, and appointed Charles Doig to design it.

According to Alfred Barnard, who visited two years later, it was one of the 'most modern of distilleries. Perhaps the best designed and most efficient of its era, great care and ingenuity having been displayed in planning.' A road was constructed to connect with the main road, there was a siding off the railway line and houses were built for the workforce and Excise officers; 20 men were recruited, mostly being paid £1 per week (the cooper was the highest paid at £1.35 per week). The distillery went into production in mid July 1897.

Production climbed until 1903, then fell back, until it was halved between 1906 and 1910. The distillery closed 1911/12, reopened 1913 and continued to prosper until 1925. With the onset of the Great Depression, it close again in 1928, and remained so until 1948. In 1950 Saladin maltings replaced the earlier floor maltings. These remain on site to this day, but unused. Tamdhu was doubled in size (to four stills) in 1972, and two more stills were added in 1975. It was mothballed in 2010, then sold to Ian Macleod & Co. the following year, who brought the distillery back into production in 2012 and launched their first core expression (a 10YO) in 2013.

Anticipating increased traffic on the Strathspey railway, its owner, the Great North of Scotland Railway Company, built a station at Tamdhu, called Dalbeallie. From 1976 to 2009 this was the distillery's visitor centre.

CURIOSITIES: In the early months of production, the spirit was found not to have the 'body' of other 'Glenlivet-style' makes. The Manager believed that the spring water was to blame and experimented with water drawn from the Knockando Burn, which, he reported, made for a 'thicker and better spirit'. He also began to use locally grown barley (most of the barley used in the early months of production was 'foreign'). Some of the partners, however, preferred the spring water make.

RAW MATERIALS: Soft water from bore-holes on site and a spring beneath the distillery. Saladin box maltings since 1950, producing lightly peated malt, from locally sourced peat. Malt from independent maltsters.

PLANT: Semi-Lauter mash tun (11 tonnes per mash). Nine Oregon pine washbacks. Three plain wash stills (10,500 litres charge), three plain spirit stills (13,000 litres charge); all stills indirect fired by steam. Shell and tube condensers.

MATURATION: Four dunnage and one racked warehouse on site, with plans to build a further six palletised warehouses. Ex-sherry, refill hogsheads and butts for single malt bottlings. American barrels for blending purposes.

STYLE: Fresh honeyed apple, with a light smokiness. Good depth.

MATURE CHARACTER: Tamdhu is a 'well-mannered' malt, well-made and displaying all the virtues of Speyside. The nose is sweet and estery, with fresh fruits (including Ogen melon), nail varnish remover and a hint of smoke. The mouthfeel is voluptuous, the taste sweet and fruity, with a trace of peat. Medium-bodied.

Tamnavulin

REGION
Speyside

ADDRESS
Ballindalloch, Moray

PHONE
01807 590285

WEBSITE
www.whyteandmackay.
co.uk

OWNER
Whyte & Mackay Ltd

VISITORS
No

CAPACITY
4.3m L.P.A.

A 1982 advertisement for Tamnavulin proclaimed: 'Fewer than one in 500 will pick up the Gaelic this year' with the strapline, 'Say "Tamnavulin" and let the malt speak for itself.'

HISTORICAL NOTES: Tamnavulin was built by the Tamnavulin-Glenlivet Distillery Company, a subsidiary of Invergordon Distillers Ltd, in 1965/66. At the time it was one of only two distilleries in Glenlivet (the other being The Glenlivet; in 1973/74 they were joined by Braeval, originally named Braes of Glenlivet). It was equipped with six stills, reflecting the demand for Scotch at that time.

It is utilitarian in design, although it enjoys an attractive situation in a steep glen carved by the River Livet. In the mid 1980s the old carding mill that gives its name to the place was converted into a visitor centre, but this is currently closed.

Whyte & Mackay bought Invergordon in 1993 and mothballed the distillery two years later. In January 2007 the company started a major refurbishment, which was completed in August the same year, by which time Whyte & Mackay had been bought by the Spirits Division of United Breweries of India (May 2007), who went into full production, the malt being almost entirely used for blending.

In May 2014, Whyte & Mackay was sold to the Philippines brandy maker Emperador (see *Dalmore*, *Invergordon*).

CURIOSITIES: *Mhuilinn* (pronounced 'voolin') is the Gaelic for 'mill'; 'Tamnavulin' is 'the mill on the hill'.

RAW MATERIALS: Unpeated malt from independent maltsters. Process water from springs in the Easterton surrounding hills; cooling water from the River Livet or Allt a' Choire.

PLANT: Stainless steel full-Lauter mash tun (10.52 tonnes per mash); four stainless, four mild steel washbacks. Three plain wash stills (18,000 litres charge); three plain spirit stills (15,000 litres charge). All indirect fired. Shell-and-tube condensers.

MATURATION: Mix of ex-sherry, ex-bourbon and refill hogsheads. Matured on site in racked warehouses.

STYLE: Light, sweet, slightly peppery and grassy.

MATURE CHARACTER: Fresh and herbal, with dried parsley, green vegetables, lemongrass, a hint of camphor, which comes through in the minty mouthfeel, with lemon meringue and camomile tea in the taste. Light-bodied.

Teaninich

REGION
Highland (North)

ADDRESS
Alness, Ross and Cromarty

PHONE
01349 882461/885001

WEBSITE
www.malts.com

OWNER
Diageo plc

VISITORS
By appointment

CAPACITY
10m L.P.A.

'Beautifully situated on the margin of the sea ... the only distillery north of Inverness that is lighted by electricity ... besides which it possesses telephonic communication.'
Alfred Barnard, 1887

HISTORICAL NOTES: The district around Alness, where stands Teaninich (pronounced 'Chee-an-in-ick'), is Munro country. It was known as Ferindonald, from the Gaelic *Fearainne Domnuill* or Donald's Land – a reference to the founder of the clan, who received lands here from Malcolm II (1005–34) for help against Norse invaders. 'Teaninich' comes from *Taigh an Aonaich* meaning 'the house on the hill'.

Captain Hugh Munro owned the Teaninich estate and built the distillery here in 1817 with the encouragement of the local lairds, who were determined to stamp out the widespread illicit distilling in the county and wanted to provide farmers with an alternative legal outlet for their barley crop. This was only partially successful: three of the four legal distilleries founded in Ross-shire at this time failed.

He was succeeded by General John Munro, who was much abroad in India and leased the site to Robert Pattison (1850), then John McGilchrist Ross (1869). General Munro was an exemplary landlord. During the 'Hungry Forties', he not only supported his poor tenants with money, but 'administers to their relief by daily personal visits, by supplying them with medicines, distributing among them meals and other provisions, and by providing them with fuel during the rigour of the winter season'. (*New Statistical Account*, 1845)

Ross gave up the tenancy in 1895 and was succeeded by Munro & Cameron, whisky brokers and spirit merchants in Elgin. They extended and refurbished Teaninich in 1899, and in 1904 Innes Cameron became sole proprietor. He died in 1932, and the following year his trustees sold to S.M.D. (D.C.L.). It was closed from 1939 to 1946, as a result of wartime restrictions on barley supply, but otherwise has been in continuous production throughout its long life (except for 1985–1990).

The stillhouse was refitted in 1962 and a second pair of stills added – a pair of 'very small stills' had

been removed in 1946 – and all four were converted to internal heating by steam. In 1970 a brand new stillhouse was built alongside, with six stills – this was a common S.M.D. practice at the time (see *Linkwood, Glendullan, Brora/Clynelish, Glenlossie/ Mannochmore*). It was simply named 'A. Side'; the milling, mashing and fermentation part of the original distillery ('B. Side') was rebuilt three years later, and a dark-grains plant built on site in 1975. At this time, Teaninich was the largest of S.M.D.'s distilleries, with a capacity of 6m L.P.A. The two sides were operated separately, and their makes vatted together prior to maturation. B Side was mothballed in 1984 and decommissioned in 1999.

In 2014/15 capacity was increased from 6 to 10 million L.P.A. by the installation of new equipment for mashing and distillation, including a larger mash conversion vessel and mash filter, ten new washbacks, the conversion of three existing wash stills into spirit stills and the installation of three new spirit stills. Plans for a brand-new distillery on an adjacent site, which would increase Teaninich's capacity to 13 million L.P.A. were also announced, but these were put on hold in October 2014, owing to a downturn in the Chinese and South American markets.

CURIOSITIES: Captain Hugh Munro was known as 'the blind captain', having lost his sight owing to an injury during the Napoleonic War, aged 24.

Innes Cameron had substantial interests in Benrinnes, Linkwood and Tamdhu Distilleries, and became Chairman of the Malt Distillers Association.

Teaninich has a unique Asnong hammer mill to grind its malt, rather than the more usual roller mill. It also employs a unique mash conversion vessel, where a vortex stirs the mash like a watery porridge. The 'porridge' then goes to a Meura filter press where it is squeezed between 24 cloth plates, and the wort collected. A second water is then added, through the filter; the remaining liquid, called 'weak worts', becomes the first water of the next batch. The filter plates are then separated to allow the draff to be collected. Each pressing takes two hours, and it takes three pressings to fill one washback. This technique is used in the brewing industry, but not by whisky distillers except Inchdairnie (see entry p.246). It was installed at Teaninich in 2000.

RAW MATERIALS: Process and cooling water from the Dairywell Spring. Unpeated malt from Glen Ord maltings.

PLANT: New (and unique) stainless steel mash conversion vessel (seven tonnes), eighteen Oregon pine washbacks, four stainless steel. Six boil-ball wash stills (18,500 litres charge), six boil-ball spirit stills (all 15,000 litres charge). All indirect fired with shell-and-tube condensers.

MATURATION: Mainly refill ex-bourbon hogsheads, some ex-sherry butts. New-make tankered to Menstrie for cask filling and maturation.

STYLE: Grassy, oily.

MATURE CHARACTER: Teaninich is a robust malt, in the North Highland style. The nose is sweet and slightly waxy, with dandelion, green leaves, green apples and gooseberry notes. The mouthfeel is smooth and mouth-filling, and the taste lightly sweet with citric hints and a thread of smoke. Medium-bodied.

Tobermory

REGION
Highland (Island)

ADDRESS
Tobermory, Isle of Mull, Argyll

PHONE
01688 302647

WEBSITE
www.tobermory distillery.com

OWNER
Distell Group Ltd

VISITORS
Visitor centre and shop

CAPACITY
1m L.P.A.

HISTORICAL NOTES: Tobermory Distillery has had a chequered history. Like Pulteney (see entry), the village itself was planned in the 1780s to be a 'fishing station' by the landlord, the fifth Duke of Argyll, with the encouragement of the British Fisheries Society. The work was organised by 'Mr Stevenson of Oban' (see *Oban*) and the Duke's chamberlain, but the plan was criticised by another Stevenson, Robert, the great lighthouse builder and father of Robert Louis, and one way or another Tobermory never became a serious fishing port.

The distillery may have been founded during this period – 1795 or 1798 – by John Sinclair Esq., 'merchant' and proprietor of Lochaline in Morvern, although he did not receive a charter to the land until 1823, and there is no sign of a distillery on the site in William Daniell's engraving of Tobermory in 1813. Barnard says, with unusual brevity, that 'the distillery was established in 1823'. It was originally named 'Ledaig'.

Sinclair seems to have been the licensee until 1837, then there is a gap until 1878, then a couple of owners, the last of whom was sequestrated in 1887, soon after which Tobermory Distillery was acquired by John Hopkins & Company, who went on to buy Speyburn Distillery in 1897, and joined D.C.L. in 1916.

S.M.D. mothballed Tobermory in 1930 (it was subsequently used as a canteen and a power station), before selling it in 1972 to Ledaig Distillery (Tobermory) Ltd, a partnership between Pedro Domecq and a Liverpool shipping company. Ledaig Distillery went into production in 1972, but by 1975 the new owner was in receivership and four years later it was sold to the Kirkleavington Property Company, of Cleckheaton in Yorkshire, which converted some of the distillery

The distillery's situation on the sheltered shore at the mouth of the fast-flowing Tobermory River met all supply requirements: water, barley, fuel, not to mention the shipping of casks of whisky.

buildings into flats in 1982, and rented others for cheese storage.

Production was resumed in 1989, then Tobermory Distillery was sold to Burn Stewart Distillers for £600,000 (plus £200,000 for stock) in 1993.

In April 2013, Burn Stewart was bought by the South African drinks giant Distell Group Ltd, for £160 million.

CURIOSITIES: Tobermory, the brand, has been used for blended Scotch, The Malt Scotch (a blended malt) and single malt whisky. The malt whisky distilled from 1972 to 1974 was bottled as Ledaig (and is highly sought after). Since Burn Stewart took over the brand names have been rationalised: Tobermory is an unpeated malt and Ledaig heavily peated. Currently, production is equally split between each style of spirit.

RAW MATERIALS: Soft peaty water from Gearr a'Bkimm lochan above the distillery. Unpeated malt from Greencore and heavily peated malt (35ppm phenols) from Port Ellen maltings.

PLANT: Cast iron rake-and-plough mash tun with a copper canopy (five tonnes per mash); four Oregon pine washbacks with no switchers. Two boil-ball wash stills (18,000 litres charge). Two boil-ball spirit stills (14,500 litres charge). Indirect fired. Shell-and-tube condensers.

MATURATION: Primarily first-fill and refill bourbon casks, with some ex-sherry casks. Small number of casks on site; rest matured at Deanston Distillery.

STYLE: Malty (Tobermory) and smoky (Ledaig).

MATURE CHARACTER: Historically, Tobermory has been variable. The 'Standard' 10YO is somewhat maritime and somewhat industrial (in the faint, oily smokiness); cereal notes abound, with light fruitiness. The mouthfeel is soft and the taste dryish, with nuts and apple brandy, and a hint of smoke. Light-bodied.

Tomatin

REGION
Highland (Central)

ADDRESS
Tomatin, Inverness-shire

PHONE
01463 2481441

WEBSITE
www.tomatin.com

OWNER
Tomatin Distillery
Company (Marubeni
Europe plc)

VISITORS
Visitor centre and gift
shop

CAPACITY
5m L.P.A.

HISTORICAL NOTES: Tomatin means 'the hillock of the juniper' (in Gaelic, *tom – aiteann*), and at over 1,000 feet this is no mean hillock! The first distillery here, just off the A9 some 18 miles south of Inverness, was built in 1897 by a group of local businessmen trading as the Tomatin Spey District Distillery Company Ltd. They went into liquidation in 1906, but the distillery was reopened by the New Tomatin Distillers Company Ltd in 1909.

The distillery was extended from two stills to four in 1956; two more were added in 1958, four in 1961, and a staggering further twelve stills in 1974, as well as a dark-grains plant. With 23 stills, Tomatin now had the greatest capacity of any malt whisky distillery – 12m L.P.A. per annum – and operated to capacity for a period in the mid 1970s, consuming 600 tonnes of malt a week. Production was dramatically cut in the 1980s, and 11 stills were removed in 1997/98.

Tomatin Distillers plc went into liquidation in 1986 and the distillery was sold to Takara Shuzo Company and Okura & Company of Japan, both long-standing customers. This was the first time a Japanese company had entered the Scotch whisky industry. Marubeni replaced Okura around 2000, and the consortium was joined by the Japanese distributors Kokubu in 2006.

Until 2012, Tomatin produced mainly for blending (not least for the company's own blends, The Antiquary and Talisman), with a core range of 12, 18 and 30YO, plus annual 'vintage' releases and single cask bottlings. Now the focus is on single malts, with a range of expressions, including (since 2013) a peated version named Cù Bòcan. The name derives from a spectral hound that roams the moors around Tomatin. Sales of Tomatin single malt have risen from 12,000 to 34,000 nine litre cases over the past ten years.

The old established brand of blended Scotch, The Antiquary, was bought by Tomatin in 1996 and bottled at 12 and 21 years. It was created by John and William Hardie, whose partnership was said to have been founded in 1857, which makes it one of the earliest blends.

CURIOSITIES: The hamlet of Tomatin stands on an old cattle drovers' road, and there was an illicit still here, beside the 'Old Laird's House', where the drovers would fill their flasks.

Tomatin was the first whisky distillery to introduce the Lauter mash tun from the German brewing industry in 1974, when the distillery's capacity was doubled. Prior to this all mash tuns were traditional infusion vessels with rake-and-plough (rack-and-pinion) stirring gear. The virtue of the Lauter tun is that its rotating arms are equipped with 'blades' which gently lift the malt bed to allow the worts to drain, rather than agitating the entire bed with rakes that revolve as well as rotate. This makes for clearer worts, especially if the mash tun is a full-Lauter, where the blades can be raised and lowered as well as rotated.

Cù Bòcan literally translates as 'the hobgoblin dog'. Tomatin's website tells us that this 'hellhound' has stalked residents of the remote Highland village for centuries, appearing about once in a generation. 'A distillery worker out late was once relentlessly pursued by an imposing black beast, steam spiralling from flared nostrils, teeth bared.'

RAW MATERIALS: Water from Allt-na-Frithe, 'the free burn'. Floor maltings until 1973. Unpeated and peated malt (2–5ppm phenols) from Simpson's and Greencore.

PLANT: One stainless steel Lauter tun (eight tonnes per mash) – also an old traditional tun in place. Twelve stainless steel washbacks (and six disused ones of cast iron). Six boil-ball wash stills (16,800 litres charge; six boil-ball spirit stills (16,800 litres charge); all indirect fired. Shell-and-tube condensers.

MATURATION: Mainly refill hogsheads; also sherry hogs, bourbon barrels and a small amount of new hogs from

Speyside Cooperage. Finish in Spanish ex-sherry for about nine months to a year. Fourteen bonded warehouses on site with a capacity of *c.*197,000 casks.

STYLE: Light, fresh, grassy, some vanilla.

MATURE CHARACTER: The nose is sweet, malty and aromatic with vegetal notes. The taste is sweet to start, with cereal flavours supported by caramelised fruits (orange especially) and assorted nuts. A very faint smokiness. Medium-bodied.

Tomintoul

REGION
Speyside

ADDRESS
Tomintoul, Moray

PHONE
01807 590274

WEBSITE
www.tomintouldistillery.
co.uk

OWNER
Angus Dundee Distillers
plc

VISITORS
By appointment

CAPACITY
3.3m L.P.A.

HISTORICAL NOTES: The functional-looking distillery was built in 1964/65 by two firms of whisky brokers and blenders in Glasgow (Hay Macleod Ltd and W. & S. Strong Ltd), who sold to Lonrho in 1973. Founded in 1909, Lonrho (the name is short for the 'London and Rhodesia Mining Company') was headed by a controversial entrepreneur, Tiny Rowland, from 1961, who transformed it into a global conglomerate. Whyte & Mackay was also acquired in 1973, and ownership of Tomintoul was transferred. Capacity doubled in 1974 (to four stills). W. & M. (which was then owned by Jim Beam) sold the distillery to the London-based blenders Angus Dundee Distillers plc in 2000, and although the majority of the spirit is used in their many blends, there has been considerably more focus on single malts. The core range (all kosher registered) are at 10, 14, 16 and 21 years, and since 2005 a range of peated expressions named Old Ballantruan, after the distillery's water source.

CURIOSITIES: Established over 50 years, Angus Dundee is a blending and bottling company owned and managed by three generations of the Hillman family. The acquisition of Tomintoul was the company's first venture into distilling. Glencadam was acquired in 2003.

Tomintoul (pronounced 'Tomintowel') is familiar to U.K. radio listeners for usually being the first village to be cut off by snow!

RAW MATERIALS: Water from Ballantruan Spring. Malt from independent maltsters, usually unpeated.

PLANT: Semi-Lauter mash tun (11.6 tonnes). Six stainless steel washbacks. Two boil-ball wash stills (15,000 litres charge); two boil-ball spirit stills (9,000 litres charge). All indirect fired by steam. Shell-and-tube condensers, one a multi-pass to create hot water for the evaporator.

MATURATION: Mainly refill hogsheads; some first-fill and some Oloroso sherry butts. Six high-racked warehouses on site (holding 114,000 casks in total). Many other sites used.

STYLE: Light, fragrant, floral, fruity. Speyside, now also with a peaty variant.

MATURE CHARACTER: The bottle describes Tomintoul as 'The Gentle Dram'. The style is light and delicate; the nose grassy, perfumed, lemony; the taste sweet, with breakfast cereal and nuts, and a shortish finish. Light-bodied.

To Tormore

REGION
Speyside

ADDRESS
Advie, Grantown-on-Spey,
Moray

PHONE
01807 510244

WEBSITE
www.tormore.com

OWNER
Chivas Brothers

VISITORS
By appointment

CAPACITY
4.4m L.P.A.

HISTORICAL NOTES: Tormore was one of several distilleries built to satisfy the post-World War II thirst for whisky, and the first to be raised from scratch on a greenfield site. It was built between 1958 and 1960 by Schenley International, U.S. agents for Dewar's, which had bought Seager Evans in 1956 (owner of Long John) and the Black Bottle blend in 1959.

The design for the distillery was commissioned from Sir Albert Richardson, past President of the Royal Academy. A contemporary described it as 'a masterpiece of distillery architecture' and whisky writer Michael Jackson compared it to 'a spa offering a mountain-water cure'!

In 1972 capacity was doubled (to eight stills), and three years later Schenley sold out to Whitbread. Whitbread's spirits division was in turn acquired by Allied Lyons in 1989, and Allied bought by Pernod Ricard/Chivas Brothers in 2005.

Tormore was closed during the winter of 2011/12 and three more washbacks were added, increasing production from 3.7 million to 4.4 million L.P.A. At the same time the stills were converted from steam coils to external heaters to save energy. Further savings were achieved in 2014 when Tormore, The Glenlivet, Cragganmore and Tomintoul Distileries were all connected by a 16-mile-long gas pipeline.

Tormore has always been a blending whisky, although small amounts have been bottled as single malt at 12YO since the early 1980s. In 2013 this was replaced by a 14YO and a 16YO.

CURIOSITIES: As late as 1983, Tormore was being labelled and promoted as The Tormore-Glenlivet: the Glenlivet appellation was still deemed to be a guarantee of quality, although most of the 28 distilleries which once adopted the suffix had by then dropped it.

The neat lawns in front of the main building are decorated with bushes clipped into a topiary of still-shapes. All but three of the smart white houses that surround the distillery, which were built for distillery workers, are now owned privately.

The conspicious distillery clock is programmed to play four different tunes per hour. Broken for many years, it was restored to life in 2007.

RAW MATERIALS: Unpeated malt from independent maltsters. Soft water from Achvockie Burn.

PLANT: Full-Lauter mash tun (10.4 tonnes); eleven stainless steel washbacks. Four plain wash stills (11,000 litres charge); four plain spirit stills (7,500 litres charge). All fitted with purifiers. All heated externally by gas. Shell-and-tube condensers.

MATURATION: Mainly refill U.S. hogsheads; some first-fill.

STYLE: Sweet, fruity and estery.

MATURE CHARACTER: Tormore's label describes it as 'The Pearl of Strathspey'; reflecting the whisky's 'brilliant appearance and honouring the freshwater pearl mussels flourishing in the pure clean waters of the River Spey'.

A firm Speyside character with a malty, nutty (almonds, coconuts) nose. A pleasant, smooth texture and a sweet, honeyed taste, drying in the relatively short finish. Medium-bodied.

Tullibardine

REGION
Highland (South)

ADDRESS
Blackford, by
Auchterarder, Perthshire

PHONE
01764 682252

WEBSITE
www.tullibardine.com

OWNER
Tullibardine Distillery Ltd
(Maison Picard)

VISITORS
Visitor centre, restaurant
and sizable retail park

CAPACITY
3m L.P.A.

HISTORICAL NOTES: The distillery was designed and built by William Delmé-Evans in 1949 (the first 'stand alone' distillery – that is, not within another distillery – to be built since 1900, presaging the boom of the 1950s and 1960s). He went on to design Macduff, Jura and Glenallachie Distilleries.

It was sold to Brodie Hepburn (whisky blenders in Glasgow) in 1953, and passed to Invergordon in 1971 (expanded to four stills in 1973/4), then to Whyte & Mackay when they bought Invergordon in 1993 (who mothballed it the next year).

In 2003 a consortium of whisky men bought the distillery and resumed production, having appointed the highly experienced John Black as Manager. Money for the project was raised by obtaining planning permission for a retail park called 'Eagle's Gate' on land adjacent to the distillery, which was then sold to a leading property developer. Since the distillery stands on the main road between Glasgow and Inverness (the A9), this was a realistic idea.

In 2011, Tullibardine was bought by the French wine-maker Maison Michel Picard, owner of the domain Chassagne-Montrachet (among others) in Burgundy and itself a family-owned company. In 2013, they launched a new range of malts and redesigned the visitor centre.

CURIOSITIES: Eagle's Gate Retail Park proved to be a mixed blessing, since it dominated the site, and in January 2015 the new owners bought the retail park and converted the shop units into maturation halls, a dry goods and cased goods store and a vatting hall and bottling line. They also acquired a stand-alone building in the car park (formerly a furniture shop) and converted it into a cooperage. These opened in April 2016, and will form part of 'connoisseurs tour'. The existing shop will be upgraded and a tasting room added in 2017.

Tullibardine is one of only six distilleries to bottle

Blackford, where Tullibardine Distillery is located, has been famous for its water for hundreds of years. The local brewery produced an ale for the coronation of King James IV in 1488, and today the famous Highland Spring water is bottled there.

on site. As well as it's own malt, single malts under the Highland Queen and Muirhead's labels, and for Costco North America (the Kirkland brand) are bottled here.

Scotland's most famous hotel, Gleneagles, is a neighbour. The first 'country club' hotel in the U.K., it was completed in 1924 and retains an old-fashioned charm, for all its state-of-the-art luxury, including the best restaurant in Scotland and comprehensive sporting facilities (including four golf courses).

In September 2012, a deal was signed with Celtic Renewables, a spin-out company from Napier University in Edinburgh, to convert draff and pot ale into butanol, which can be used to fuel road vehicles.

RAW MATERIALS: Soft water from the Danny Burn, with its source in the Ochil Hills, where the brewery water came from. Unpeated malt from Greencore Maltings, Carnoustie.

PLANT: Stainless steel full-Lauter mash tun with a peaked canopy (six tonnes), nine stainless steel washbacks. Two plain wash stills (15,000 litres charge) and two lamp-glass spirit stills (11,000 litres charge). All indirect fired. Shell-and-tube condensers.

MATURATION: Predominantly ex-bourbon barrels, with some Oloroso sherry and Pedro Ximénez hogsheads and butts. Racked warehouse on site.

STYLE: Sweet, fruity and malty.

MATURE CHARACTER: Tullibardine is remarkably consistent in its make. The key flavour is malt, balanced by a zesty fruitiness. The nose is all of this, and some peach and melon notes. The taste is sweet to start, then drying, with biscuits and light caramel in the middle. Medium-bodied.

Wolfburn

REGION
Highland (North)

ADDRESS
Thurso, Caithness

PHONE
01847 891051

WEBSITE
www.wolfburn.com

OWNER
Aurora Brewing Ltd

VISITORS
Visitors by arrangement

CAPACITY
132,000 L.P.A.

HISTORICAL NOTES: The site chosen by the founders of this distillery is close to the site of a former Wolfburn Distillery (1821 to mid 1850s). It is the most northern distillery on the Scottish mainland.

Building commenced in August 2012 and the first spirit flowed on St Andrew's Day (25 January) 2013. By October Wolfburn had produced 100,000 litres of spirit, of a light and fragrant style; in 2014 a lightly peated spirit (10ppm phenols) was produced for two months, increased to four months in 2015. Shane Fraser, who was formerly Manager at Glenfarclas, is production manager.

CURIOSITIES: Wolfburn is situated in the Flow Country, a vast area of wetlands and blanket peat covering around 4,000 square kilometres, the largest such area in Europe, possibly in the world, and a potential UNESCO World Heritage site on account of the wildlife it supports.

The Wolfburn logo is taken from a drawing by Konrad Gesner, the 16th-century physician and naturalist, and appears in his work *The History of Four-footed Beasts and Serpents*. In Gesner's day, the wolf was a common sight in the far north of Scotland, and on the coast it was said to have a supernatural relative: the sea-wolf. According to the lore of the time, the sea-wolf did 'liveth both on sea and land'. Gesner's woodcut is thought to show the creature walking on water, but this is not the limit of its gifts: the sea-wolf is also said to bring good luck to all those fortunate enough to see it.

RAW MATERIALS: 90% unpeated malt; 10% of annual production is done with lightly peated (10ppm) malt. (A limited release peated edition is expected annually, commencing 2017.) Soft water from the Wolf Burn, which rises in the Flow Country, fed by springs.

PLANT: Semi-Lauter mash tun (1.1 tonnes), with a copper canopy. Three stainless steel washbacks. One boil-ball spirit still (5,500 litres charge), one plain wash still (3,600 litres charge).

MATURATION: Approximately one third in ex-bourbon quarter casks, one third in ex-bourbon hogsheads and barrels, and one third in ex-sherry butts.

STYLE: Light, sweet and malty, with notes of dried apricot, banana and light spice.

MATURE CHARACTER: The first bottling of Wolfburn was released on 4 March 2016. It has a distinctly 'northern' flavour profile. Sweet and slightly salty (maritime), with a distinct smokiness, which may be owing to the fact that it has been matured in ex-Laphroaig quarter casks. All the spirit is matured on site in a mix of butts (American and European oak), barrels and quarter casks.

Facts and Figures

Who Owns Whom?

Including a brief account of recent changes in ownership.
(Earlier histories appear under individual entries.)

ABHAINN DEARG (UK PRIVATE)
Malt distilleries: Abhainn Dearg

ANNANDALE DISTILLERY COMPANY (UK PRIVATE)

ARBIKIE ESTATE LTD (UK PRIVATE)
Malt and grain whiskies

ARDNAMURCHAN DISTILLERY LTD (UK PRIVATE)

BALLINDALLOCH ESTATE (UK PRIVATE)

BEAM SUNTORY

Beam Inc. is the latest incarnation of the owners of Jim Beam, the world's best-selling bourbon. Until 2011 the company was named Beam Global and was itself owned (since 1969) by American Brands (now named Fortune Brands), which bought Whyte & Mackay in 1990, changed its name to J.B.B. (Greater Europe) and sold to a management buy-out in 2001. The new owner was named Kyndal Spirits Ltd., but changed its name back to Whyte & Mackay the following year.

Fortune Brands re-entered the Scotch whisky industry with the surprise purchase of William Teacher & Sons and its associated distilleries when Allied Domecq was broken up in 2005 (see Chivas Brothers below).

Beam Global split from Fortune Brands in 2011; the latter was dissolved next year, and Beam Global became simply 'Beam Inc.'. This company was bought by Suntory in April 2014 to become Beam Suntory Inc., the world's number three spirits company, with its HQ in Deerfield, Illinois. Suntory, the leading Japanese distiller, entered the Scotch whisky industry with the purchase of 35% of Morrison Bowmore in 1985, acquiring whole ownership in 1994. Morrison Bowmore disappeared as a trading entity at New Year 2015.

Malt distilleries: Ardmore, Laphroaig, Bowmore, Auchentoshan, Glengarioch

BEN NEVIS DISTILLERY LTD (NIKKA, ASAHI BREWERIES)

Malt distilleries: Ben Nevis

BENRIACH-GLENDRONACH DISTILLERS LTD (UK PRIVATE)

Malt distilleries: BenRiach, Glendronach, Glenglassaugh

BLADNOCH DISTILLERY LTD (AUSTRALIA PRIVATE)

Malt distilleries: Bladnoch

BRUICHLADDICH DISTILLERY LTD (RÉMY COINTREAU)

Malt distilleries: Bruichladdich

BURN STEWART DISTILLERS LTD (DISTELL GROUP, S. AFRICA)

Burn Stewart was acquired in 2002 by Trinidad-based C.L. World Brands, owners of Angostura Bitters, Hine Cognac and Rhum Rum.

In April 2013, C.L. World Brands sold Burn Stewart and its distilleries to the Distell Group of South Africa – Africa's leading spirits, wines and ready-to-drink business – for £160 million.

Malt distilleries: Deanston, Tobermory, Bunnahabhain

CHIVAS BROTHERS LTD (PERNOD RICARD)

Owned by Pernod Ricard since 2001, Chivas Brothers had been owned by the Canadian distilling giant Seagram since 1949. Seagram bought Strathisla Distillery the next year and built Glen Keith in 1957, Braeval in 1973 and Allt-a-Bhainne in 1975. In 1978, The Seagram Company acquired The Glenlivet Distillers (owning The Glenlivet, Glen Grant, Caperdonich, Longmorn and BenRiach Distilleries), but decided to divest itself of its massive interest in alcoholic beverages in 2001, when its assets were divided between Pernod Ricard and Diageo.

In 2005, Pernod Ricard went on to acquire the majority of Allied Domecq's distilleries (Glenburgie, Glentauchers, Imperial, Miltonduff, Tormore, Scapa) and its leading blended whisky, Ballantine's, selling Glen Grant to satisfy the Monopolies and Mergers Commission.

Until these developments, Pernod Ricard had been a small player in the Scotch whisky industry, through its subsidiary Campbell Distillers (owning Aberlour, Glenallachie and Edradour Distilleries – the latter later sold). Now it is the second largest distiller in the world.

Grain distilleries: Strathclyde

Malt distilleries: Aberlour, Allt-a-Bhainne, Braeval, Glenburgie, Glenallachie, Glen Keith, The Glenlivet, Glentauchers, Miltonduff, Longmorn, Tormore, Scapa, Strathislu, Dalmunach

DAFTMILL (UK PRIVATE)

JOHN DEWAR & SONS LTD (BACARDI PLC)

Bacardi, the family-owned rum distiller, entered the Scotch whisky industry when it bought William Lawson Distillers Ltd and Macduff Distillery in 1992.

When Grand Metropolitan (Independent Distillers & Vintners) merged with Guinness (United Distillers) in 1997, the Monopolies and Mergers Commission insisted that they divest themselves of a major brand. As a result, John Dewar & Sons, together with four distilleries, was sold to Bacardi.

Malt distilleries: Aberfeldy, Aultmore, Craigellachie, Macduff, Royal Brackla

ANGUS DUNDEE LTD

Malt distilleries: Glencadam, Tomintoul

DIAGEO PLC

Diageo was formed as a holding company in 1998, following the merger of Grand Metropolitan and Guinness the previous year, with United Distillers & Vintners (U.D.V.) as its alcoholic beverages subsidiary, combining the previous operational divisions, United Distillers (U.D.) and International Distillers & Vintners (I.D.V.). The name 'Diageo' comes from 'dia' (Latin for 'day') and 'geo' (Greek for 'world'), suggesting that the company's brands are sold everyday, all over the world. In 2000, the corporate structure was simplified, so that U.D.V. was replaced by Diageo as the trading entity, and in 2002 U.D.V. ceased to exist.

United Distillers came about in 1987, following the acquisition by Guinness of the leading Scotch whisky company The Distillers Company Limited (D.C.L.).

Grain distilleries: Cameronbridge, 50% of North British

Malt distilleries: Auchroisk, Benrinnes, Blair Athol, Caol Ila, Cardhu, Clynelish, Cragganmore, Dailuaine, Dalwhinnie, Dufftown, Glendullan, Glen Elgin, Glenkinchie, Glenlossie, Glen Ord, Glen Spey, Inchgower, Knockando, Lagavulin, Linkwood, Mannochmore, Mortlach, Oban, Roseisle, Royal Lochnagar, Strathmill, Talisker, Teaninich

EDEN MILL (UK PRIVATE)

THE EDRINGTON GROUP LTD

Another name change! Edrington was formerly called Highland Distilleries (established 1887) – this became a public limited company in 1982, then withdrew from public listing in 1998, when the name was altered to Highland Distillers Ltd. The company had always been closely associated with Robertson & Baxter (Glasgow whisky brokers and blenders, established 1861). In 1944, ownership of R.&B. (and the complex relationship with Highland) passed to the three grand-daughters of the founder. To avoid punitive death duties and secure the independence of the firm, in 1961 the sisters transferred all their capital to a holding company, Edrington Ltd (named after a family farm in Berwickshire), then gifted the shares they received in exchange to a charitable trust, The Robertson Trust – now one of Scotland's leading charitable institutions. Edrington Ltd became The Edrington Group in 1999, when it took over Highland Distillers Ltd.

Grain distilleries: 50% of North British

Malt distilleries: Glenrothes, Glenturret, Highland Park, 70% of The Macallan

THE GLENMORANGIE COMPANY (LOUIS VUITTON MOËT HENNESSY)

Glenmorangie Distillery was gradually acquired by its largest customer, Macdonald & Muir, blenders in Leith (established 1893) between 1918 and 1930. By the late 1990s M. & M. had adopted the name of their acquisition, and Glenmorangie had become a public company.

This was bought by the French luxury goods company Louis Vuitton Moët Hennessy (L.V.M.H.) in 2004, and taken out of public ownership.

Malt distilleries: Ardbeg, Glenmorangie

GLASGOW DISTILLERY COMPANY (UK PRIVATE)

GORDON & MACPHAIL

Malt distilleries: Benromach

J. & G. GRANT

Malt distilleries: Glenfarclas

WILLIAM GRANT & SONS LTD

Grain distilleries: Girvan

Malt distilleries: Ailsa Bay, Balvenie, Kininvie, Glenfiddich

INCHDAIRNIE DISTILLERY LTD (UK PRIVATE)

Malt and grain whiskies

INVER HOUSE DISTILLERS LTD (INTERBEV, THAILAND)

Inver House was founded in 1964 as a subsidiary of the American company Publicker Industries, and was bought out by its Directors in 1988. In 2001 Inver House Distillers Ltd was bought by Pacific Spirits, a subsidiary of ThaiBev plc, Southeast Asia's largest alcoholic beverage company, and integrated into that company's international division, International Beverage Holdings Ltd (InterBev), in 2006.

Malt distilleries: Balblair, Balmenach, Knockdhu, Pulteney, Speyburn

ISLE OF ARRAN DISTILLERS LTD
Malt distilleries: Arran

ISLE OF HARRIS DISTILLERS LTD (UK PRIVATE)

KILCHOMAN DISTILLERY COMPANY LTD
Malt distilleries: Kilchoman

KINGSBARNS (UK PRIVATE)

LE MARTINIQUAISE
Grain distilleries: Starlaw
Malt distilleries: Glen Moray

LOCH LOMOND DISTILLERY LTD (EXPONENT PRIVATE EQUITY)
Grain distilleries: Loch Lomond
Malt distilleries: Glen Scotia, Loch Lomond

IAN MACLEOD & COMPANY LTD
Malt distilleries: Glengoyne, Tamdhu

J. & A. MITCHELL
Malt distilleries: Springbank, Glengyle

SPEYSIDE DISTILLERS COMPANY LTD (TAIWAN PRIVATE)
Malt distilleries: Speyside

STRATHEARN DISTILLERY COMPANY (UK PRIVATE)

TOMATIN DISTILLERY COMPANY LTD (MARUBENI EUROPE PLC)
Malt distilleries: Tomatin

TULLIBARDINE DISTILLERY COMPANY LTD (MAISON MICHEL PICARD)
Malt distilleries: Tullibardine

WHYTE & MACKAY LTD (EMPERADOR, PHILIPPINES)

In 1881, James Whyte and Charles Mackay bought the wine and spirits company they had been working for, Allan & Poynter of Glasgow (established 1843). One of the conditions of sale was a change of name. The company was acquired by Sir Hugh Fraser's S.U.I.T.S. group in 1971, which was acquired by Lonhro ten years later. W. & M. was sold to Brent Walker in 1988, then, in 1990, to American Brands (which became Fortune Brands – see Beam Suntory above). In 2001, Vivian Imerman led a management buy-out, backed by a German bank; four years later he bought out the bank, and in 2007 sold the company to the United Spirits Division of the United Breweries Group, the largest brewer and distiller in India.

In 2012, it was announced that United Spirits was to be sold to Diageo. The deal was completed in 2014, but Diageo was obliged by the Office of Fair Trading to immediately sell Whyte & Mackay. The company, with its distilleries and brands, was bought by Emperador, the Philippines-based brandy distiller, for £430 million in 2014.

Grain distilleries: Invergordon

Malt distilleries: Dalmore, Fettercairn, Jura, Tamnavulin

WOLFBURN DISTILLERY LTD (UK PRIVATE)

World Consumption of Bottled Malt Whisky 2000–2013

2000	15,470,000 (litres of pure alcohol)
2001	14,550,000
2002	16,030,000
2003	16,950,000
2004	18,960,000
2005	20,500,000
2006	24,480,000
2007	23,170,000
2008	23,418,228
2009	20,875,950

2010	27,046,057	Bottled Single Malt	
	7,776,996	Bottled Blended Malt	= 34,823,053
2011	28,053,036	Bottled Single Malt	
	7,040,671	Bottled Blended Malt	= 35,093,707
2012	27,165,480	Bottled Single Malt	
	6,285,440	Bottled Blended Malt	= 33,450,920
2013	26,508,442	Bottled Single Malt	
	7,801,382	Bottled Blended Malt	= 34,309,824

Source: *Scotch Whisky Association. Following a change in how the figures are recorded, 2010 was the first time the S.W.A. was able to separate Single Malt and Blended Malt from the 'Malt Whisky' category figure.*

Top 10 Malt Whisky Markets by Volume in 2012, 2013 and 2014

These figures combine bottled single malt and bottled blended malt, and show units in thousands of litres.

	2014	2013	2012
U.S.A.	15,165	14,265	12,636
France	7,992	7,686	7,398
Taiwan	7,758	7,704	7,304
U.K.	7,191	7,047	7,074
Germany	3,750	3,537	3,366
Canada	2,592	2,421	2,286
Italy	1,872	1,935	2,052
Japan	1,809	1,674	1,611
Russia	1,359	1,206	1,170
Australia	1,305	1,161	1,080

Source: The Scotch Whisky Industry Review 2015.

Top 15 Malt Whisky Brands

in 2012, 2013 and 2014

These figures are in thousands of nine-litre cases.

	2014	2013	2012
The Glenlivet	1,080	987	917
Glenfiddich	1,052	1,051	1,036
Macallan	787	780	716
Glenmorangie	495	475	367
*The Singleton	410	345	279.5
Aberlour	287	267	222
Laphroaig	277.4	229.5	204
Balvenie	246.7	225	216
Glen Grant	221.8	209.3	232.2
Talisker	192	167	136
Cardhu	190	184	181.7
Glenfarclas	180	60	53
Bowmore	172.3	174	172.8
Lagavulin	142	133	122.1
Jura	135	148.5	112.4

* 'The Singleton' comprises malts from Glen Ord, Glendullan and Dufftown with a similar flavour profile, selling respectively in Asia, North America and Europe.

Source: The Scotch Whisky Industry Review 2015.

Top 15 Blended Whisky Brands

in 2012, 2013 and 2014

These figures are in millions of nine-litre cases.

	2014	2013	2012	Owner
Johnnie Walker Red	11.30	11.61	11.88	Diageo
Johnnie Walker Black	6.53	6.65	6.76	Diageo
Ballantine's Finest	5.43	5.29	5.03	Chivas Brothers
William Grant's	4.38	4.55	4.46	Wm. Grant
Chivas Regal	4.13	4.19	4.32	Chivas Brothers
J & B Rare	4.01	3.94	4.10	Diageo
William Lawson's	3.07	2.74	2.76	Bacardi
The Famous Grouse	3.06	2.98	3.04	Edrington
Label 5	2.47	2.51	2.50	La Martiniquaise
Dewar's White Label	2.38	2.56	2.69	Bacardi
Bell's Original	2.21	2.36	2.35	Diageo
Teacher's	2.05	2.02	2.10	Beam Suntory
Passport	1.70	1.49	1.36	Chivas Brothers
Buchanan's	1.68	1.68	1.85	Diageo
White Horse	1.68	1.51	1.32	Diageo

Source: The Scotch Whisky Industry Review 2015.

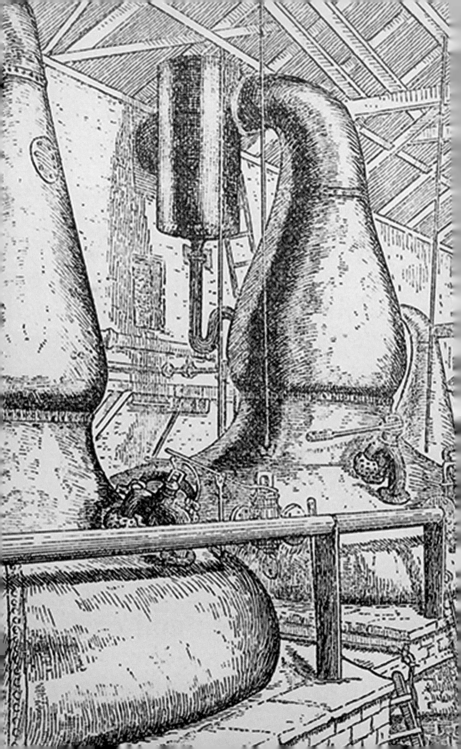

Leading Independent Bottlers

The 18 companies listed below are probably the best known of the independents. All are based in the U.K. I have not listed those which are principally distillers and blenders (like Ian Macleod & Company and Angus Dundee & Company).

This article first appeared in 'The Malt Whisky Yearbook 2008' (MagDig Media). As well as providing a thumbnail sketch of each company, Ingvar Ronde, the editor of the publication, asked me to enquire of some of the leading firms how they viewed the future for independent bottlers.

ADELPHI DISTILLERY

www.adelphidistillery.com

The Adelphi Distillery stood in the Gorbals district of Glasgow, on the site of what is now the Glasgow mosque. It was one of the largest whisky distilleries in Scotland, with both pot and patent stills. It ceased production in 1902, although the warehouses continued to be used until the 1960s.

The name was revived in 1992 by the great-grandson of the last owner, Jamie Walker, who had served an apprenticeship with Inver House Distillers and who established his company as an independent bottler of single cask single malts. He sold to two lairds from Argyllshire, who recruited Alex Bruce (formerly of J & B) as Marketing Director and, since 2015, Managing Director. The company built Ardnamurchan Distillery, which opened in 2014.

'Yes, buying casks at their optimum maturity is becoming harder, especially when we are only selecting the top 5–10% of those on offer.

However, by cementing good supply relationships, and implementing a flexible approach to labelling, wood management, and future diversification, we firmly believe we can continue to offer an exceptional range of single cask bottlings to our customers and, at the same time, promote the quality and vast spectrum of Scotch Whisky as a whole. We are, if you like, an

ambassador for the industry, a role that the independent bottler both is able, and can be proud, to perform.'
Alex Bruce, Marketing Director

BERRY BROTHERS & RUDD

www.bbr.com

The company traces its origins to 1698, and its venerable shop occupies the same site, although it was rebuilt in 1734. Like many whisky companies, the firm began as general grocers. By the 1820s it was listing wines and spirits, and storing its own casks, and during the latter half of the nineteenth century the firm concentrated increasingly on this side of the business, allowing the grocery side to dwindle.

Berry Brothers' entry into the wider world of whisky came in 1923, with the introduction of Cutty Sark. Arising out of this, the firm also acquired the exclusive right to the Glenrothes brand name, although the distillery continued to be owned by Highland (now Edrington). Berry Brothers has long offered its own malts, but only to private customers until 2002, when it was resolved to lay down more casks and expand the Berry's Own Selection range. Doug McIvor was appointed Spirits Manager: his and the company's guiding policy is simple: 'We only bottle what we enjoy drinking ourselves.'

Currently Berry Brothers offer between 25 and 30 expressions.

'We've been bottling whisky for over a hundred years and we hope still to be offering our customers a diverse selection of single malts a hundred years from now.'
Doug McIvor, Spirits Manager

BLACKADDER INTERNATIONAL

www.blackadder.com

Blackadder was established in 1995 by John Lamond and Robin Tucek (authors of the much reprinted *The Malt*

Whisky File, 1995). Robin now runs the business from Sweden, where he lives. He bottles around 100 casks of malt whisky a year in his Raw Cask, Aberdeen Distillers, Clydesdale Original and Caledonian Connections ranges. None of his whiskies is chill-filtered or coloured; many are bottled at cask strength. They are sold through specialist shops and bars in Europe, Japan and North America.

WILLIAM CADENHEAD & COMPANY

www.wmcadenhead.com

Established in Aberdeen in 1842, Cadenhead is Scotland's oldest firm of independent bottlers. William Cadenhead, brother-in-law of the founder, who took over the business in 1858, was himself a well-known poet. The business remained in his family until 1972, when it was bought by J. & A. Mitchell, owner of Springbank Distillery. It is now based in Campbeltown, with shops in Edinburgh's Old Town and in Covent Garden, London. Bottlings are at both standard and natural strength; the whisky is non chill-filtered and un-tinted. Cadenhead's offers a range of Scotch malts and also bottles non-Scotch whiskies, rums, cognacs and gins.

COMPASS BOX WHISKY COMPANY

www.compassboxwhisky.com

Compass Box was founded in 2000 by former Diageo Marketing Director (Premium Malts), John Glaser. His goal is summed up by the firm's website description: 'Creators of hand-crafted, small batch Scotch whiskies in distinctive and original styles.'

Winner of *Whisky Magazine*'s Innovator of the Year award so many times, it is embarrassing. Passionate and idiosyncratic, Compass Box offers a beautifully crafted and packaged range of blended malts (mostly using only two malts), single grains, blended whiskies and an orange-infused whisky. The company also has an excellent website at www.compassboxwhisky.com.

'First, the company's essence is to explore new flavours and create new whiskies, so we are more flexible than traditional independent bottlers of single malts. Second, we don't purvey single malts by name, and I think this will make it easier for us to obtain quality whisky as producers become more concerned with guarding their trademarks.'

John Glaser, Managing Director

DEWAR RATTRAY

www.dewarrattray.com

Founded by Andrew Dewar and William Rattray in 1868, A. Dewar Rattray began trading as an importer of French wines, Italian spirits and olive oil. By the end of the century, the firm was representing a number of well-known Highland malt distilleries – most notably Stronachie.

In 2004, the company was revived by Tim Morrison, previously of Morrison Bowmore Distillers and fourth-generation descendent of Andrew Dewar, with a view to bottling single cask malts from different regions in Scotland.

GORDON & MACPHAIL

www.gordonandmacphail.com

Established by James Gordon and John Alexander MacPhail in 1895, in the same Elgin premises it occupies today. James Gordon had already completed an apprenticeship in selecting and blending teas, wine and whisky, and from the outset he was responsible for buying casks from neighbouring distilleries. Within a year, he was joined by John Urquhart.

G. & M. have been filling their own casks since 1900, and in 1956, when John Urquhart's son, George, became Senior Partner, he took the unprecedented step of launching a range of single malts under the Connoisseurs' Choice label. Within a few years he was offering these malts for sale in Italy and France – a pioneering step.

Just as he had joined his father, so George was joined by his own children – Ian, David, Michael and Rosemary. By 1991 the firm had a staff of over 100 and new administrative offices were built adjacent to their bonded warehouses on Boroughbriggs Road, Elgin. In 1993 George realised a lifetime's ambition: to own a distillery. He bought Benromach, spent five years refurbishing and returned it to production in 1998.

George Urquhart died in 2002, but the company he developed from a small local business to global status remains a family concern, under the capable management of his children and grand-children.

G. & M. list over 300 own bottlings (mainly at 40%, some at 43%, some at cask strength) and hold a stock of over 800 whiskies (including blends) in their shop.

'Our relationship with distilleries goes back over 100 years, and since we fill our own casks, we can ensure quality, and don't have to buy from brokers. As well as young stock, we hold stocks of mature whisky going back to the 1930s – enough to look after our needs for the foreseeable future!'
Ian Chapman, Marketing Director

HART BROTHERS

www.hartbrothers.co.uk

Hart Brothers can trace their origins in the licensed trade back to the late nineteenth century when the family were licensed victualers and publicans in Paisley. However, it was not until 1964 that brothers Iain and Donald Hart incorporated the company as Wine & Spirit Wholesalers and Scotch Whisky Blenders.

Alistair Hart joined his brother Donald in the company in 1975 from Whyte & Mackay and it is his responsibility as chief blender to source mature casks of single malt whisky, which malt whiskies are further matured and only are selected for single cask bottling after careful assessment. Alistair and Donald are now assisted by their sons, Jonathan and Andrew.

HUNTER LAING & CO.

www.hunterlaing.com

After nearly 50 years in the whisky industry, Stewart Laing

left the family company, Douglas Laing & Co. (see below), to establish a new business with his sons, Scott and Andrew. He took two Douglas Laing brands with him and a stock of whisky.

Old Malt Cask is a range of single casks introduced in 1998 (bottled without chill-filtration or tinting, usually at 50% vol). Old & Rare is a super-deluxe range of single cask single malts of exceptional quality, bottled at cask strength and presented in fancy wooden chests. In addition to these well-established brands, Hunter Laing have introduced a range of younger single cask bottlings (between 10 and 20YO) named Douglas of Drumlanrig, after the Duke of Buccleuch, Chief of the Douglas family, whose seat is Drumlanrig Castle and whose signature is on every label.

In September 2014 the company bought a 35,000-square-foot warehouse in South Lanarkshire to accommodate 15,000 casks in addition to their Carron Bond nearby. Stewart Laing remarked to me, 'At the moment our whisky is stored in eighty-seven different locations.'

DOUGLAS LAING & COMPANY LTD

www.douglaslaing.com

The eponymous Frederick Douglas Laing founded the business in Glasgow in 1948, laying down casks, blending and bottling a range of whiskies. For the past 30 years the company has been managed by the founder's sons, Fred and Stewart, but in May 2013 Stewart split off to found his own business, Hunter Laing (see above), leaving Fred with the original. He immediately brought his daughter, Cara, into the company as head of brands marketing and innovation. She had previously worked with Morrison Bowmore and Whyte & Mackay.

Currently the company's leading brands are: Big Peat (a quirky Islay blended malt, which has done very well since launch), Director's Cut, Old Particular, Premier Barrel and Provenance (single malts) and McGibbons, King of Scots and The Epicurean (blends).

'Remember, we come from a blending background, specialising in super-premium blends for Far East markets. We came into the single malt category when that market dipped in the late 1990s, to accommodate stock surpluses. We still have a large reserve of mature stock – much of it built up by our ongoing fillings programmes – enough to support our blending and single malt requirements, with stock left over for exchanges.'

Fred Laing, Managing Director

JAMES MACARTHUR & COMPANY LTD

www.james-macarthur.co.uk

Founded in 1982 'with the aim of selecting whiskies from Scottish Distilleries which were not well known or had ceased to operate and bottling them at cask strength', according to their website. Sounds familiar, but 1982 was relatively early to have such ambitions. The company currently lists 30 single cask, cask-strength malts.

MURRAY MCDAVID

www.murray-mcdavid.com

Mark Reynier, Gordon Wright and Simon Coughlin, wine merchants in London, established Murray McDavid – the name comes from two of Mark's grandparents – in 1995, to offer three highly selective ranges of single malts: the Mission Range (unusual aged stock), the Celtic Heartlands range (old stock from the 1960s; now sold out) and the Murray McDavid range. The latter employed an 'additional cask evolution' (ACE) programme, a wood-finishing process which Mark described as 'an exploration of the attributes of French oak combined with American oak'.

   In 2000 the company bought Bruichladdich Distillery (see entry), and in May 2013 sold Murray McDavid to the well established whisky brokers Aceo, who went on to buy the extensive warehousing at Coleburn Distillery (see entry) to hold their own and Murray McDavid's stock and offer cask management services. Murray McDavid bottlings returned to the market in July 2014.

'We all know of the crucial role that good wood plays in creating superlative whisky; we also know that supplies of such wood are becoming more difficult to find. Our acquisition of the highly regarded Murray McDavid brand and stock demanded that we find appropriate warehousing to continue Mark Reynier's exploration of wood types, so we bought the warehousing at Coleburn. It also demanded that we secure supplies of first rate casks – ex-wine barrels, port pipes and rum puncheons, as well as traditional ex-bourbon and ex-sherry casks – and this we have also done.'
Edward Odim, Managing Director

SCOTCH MALT WHISKY SOCIETY

www.smws.co.uk

The Scotch Malt Whisky Society is a club, established in 1983, and makes its single cask bottlings available to its members at full-strength, without chill-filtration or tinting. It has its base in Leith, Edinburgh, in the oldest commercial building in Scotland, the ancient Vaults, with branch clubhouses at Queen Street, Edinburgh, and in Hatton Garden, London, as well as franchised branches in Australia, Austria, France, Italy, Japan, Netherlands, Belgium, Luxembourg, Sweden, Switzerland and the U.S.A. The Society was taken over by Glenmorangie in 2003, on the recommendation of its Board of Directors, in order to secure supplies of mature malt whisky, but remains a separate entity. Currently it bottles 150–200 casks per annum. The Society was again sold in 2015 to a group of private investors based in Edinburgh.

'Because we are part of the Glenmorangie Company, the Scotch Malt Whisky Society continues to be in a good position to secure a wide variety of interesting malts in the years to come.'
Kai Ivalo, Director

SCOTT'S SELECTION

www.speysidedistillery.co.uk

The eponymous Robert Scott worked in distilleries for 50

years, finishing as Manager of Bladnoch and then Speyside Distilleries. On his retirement in 2001, he was persuaded by George Christie, Chairman of Speyside, to put his huge experience of evaluating whisky to good use by selecting and bottling (at cask strength, without chill-filtration or tinting) a range of single casks. The casks are selected by Sandy Jamieson, the manager of Speyside Distillery and John MacDonaugh, the company's Managing Director.

SIGNATORY VINTAGE SCOTCH WHISKY COMPANY LTD

www.edradour.com

Signatory was founded in 1998 by Andrew Symington and soon won a high reputation for its unusual single cask bottlings. Typically some 50 different single malt expressions are available at any one time. Whiskies are bottled across a number of ranges, including the Un-chill Filtered Collection, the Cask Strength Collection and the Single Grain Collection. The company's original base was in Leith, the Port of Edinburgh, but was moved to Edradour in 2002 when Andrew acquired the distillery from Pernod Ricard. Extensive stocks of mature whisky are held and bottled here.

'The tree will be severely shaken, and a lot of dead wood will drop out. I realised this some years ago and invested heavily in mature whisky. As a result we have considerable stocks to use ourselves or swap with producers – certainly enough to cover ourselves for the next decade.'
Andrew Symington, Managing Director

DUNCAN TAYLOR & COMPANY LTD

www.duncantaylor.com

Duncan Taylor & Company began as whisky brokers and blenders in Glasgow in 1938, with a focus on exporting their own blended whiskies to the U.S.A. They have been filling their own casks and laying down whiskies since the 1960s, and when Euan Shand bought the company and its stocks in 2001, he acquired 'one of the largest privately held collections of rare Scotch whisky casks in the world'. Operations were moved to Huntly, Aberdeenshire, where the company also bottles its whiskies. The range includes Rarest of the Rare (single cask, cask-strength whiskies of great age from demolished distilleries), Duncan Taylor Collection (single cask, cask-strength malts and grains aged 17–42 years), N.C.2 (mainly single casks, 12–17 years, non chill-filtered at 46%), Battlehill (younger malts at 43%), Lonach (vattings of two casks from same distillery at same age, usually over 30 years to bring them up to a natural strength of over 40%), and Big Smoke (Islay whiskies at 40% and 60%). The company has won three out of eight *Whisky Magazine* Independent Bottler of the Year Awards.

Duncan Taylor obtained planning permission to build a new distillery in Huntly in May 2007, but the project is currently on hold.

'Sure, if the producers refuse to sell us new fillings, it will be difficult to operate as an independent bottler. Having said this, we have bought a lot of mature whisky over the past few years, and we have maturing stock sufficient to last many years.'
Euan Shand, Managing Director

THE VINTAGE MALT WHISKY COMPANY LTD

www.vintagemaltwhisky.com

An independent family company established in 1992 by Brian Crook, formerly Export Sales Manager with Morrison Bowmore, Vintage Malt offers a small range of single malts under its own names – Finlaggan, Tantallan, Glenalmond, etc. – and some under the distillery's own name in the Cooper's Choice range.

WEMYSS VINTAGE MALTS

www.wemyssmalts.com

A family-owned company based in Edinburgh and Fife, Wemyss (pronounced 'Weems') Vintage Malts was established in 2005. The family also own vineyards in

France and Australia, and tea gardens in Africa, and their connection with whisky goes back to the early nineteenth century, when John Haig was their tenant at Cameronbridge (see entry).

The firm currently offers non chill-filtered single cask bottlings at a minimum of 46% from Speyside, Islay, Highland and Campbeltown, identified not by the distillery name, but with reference to what the whisky tastes like (e.g. 'Spiced Figs', 'Bonfire Embers', 'Toffee Apples'), and a range of younger blended malts at 40%, also non chill-filtered.

The Wemyss family bought Kingsbarns Distillery (see entry) in 2013.

WILSON & MORGAN LTD

www.wilsonandmorgan.com

One of the few non-Scottish independent bottlers of single cask malt Scotch, Wilson & Morgan was founded in 1992 by Fabio Rossi, whose family firm (wine and olive oil merchants in Venice and later Treviso) had been involved with importing Scotch since the 1960s. The foundations of his stock were laid in the early 1990s, when he toured Scotland for the first time, laying down casks, following his own guidelines for selection: trust your palate and your instinct. Many of the single cask malts offered in his Barrel Selection range are wine-finished; they are bottled at 46% and cask strength, without chill-filtration or tinting.

THE WHISKY DISTILLERIES

OF THE

UNITED KINGDOM

THE MANUFACTURE

OF

WHISKY AND PLAIN SPIRIT.

J A Nettleton

Sources and Acknowledgements

PUBLICATIONS

The first person to write in detail about the distilleries of Scotland was Alfred Barnard in *The Whisky Distilleries of the United Kingdom* (Harpers, 1887). His book was reprinted by David & Charles in 1969 and 1987 (the latter with an introduction by Professor Michael Moss), and by Birlinn Ltd, Edinburgh, in 2003 (with an introduction by Richard Joynson). This substantial, engaging book is the bible of 'distillery baggers'.

In 1987, the centenary of Barnard's magisterial publication and a time when the whisky industry was in some difficulties, the original publisher commissioned a contemporary version from Philip Morrice (author of the useful *Schweppes Guide to Scotch*): *The Whisky Distilleries of Scotland and Ireland* (Harpers, 1987). Before this appeared Brian Spiller, the Distillers Company archivist, had written short and scholarly accounts of that company's 50 distilleries, published as individual pamphlets (*D.C.L. Distillery History Series*, 1981–83).

Among more recent publications, I was glad to have Misako Udo's *The Scottish Whisky Distilleries* (Black & White Publishing, Edinburgh, 2006) to hand, and the invaluable *Malt Whisky Yearbook 2014* (edited by Invar Ronde, Magdig Media, Shrewsbury, 2013). Gordon & MacPhail's *Distillery Profiles* in their periodical *Scotch Whisky News* were useful, as was Ulf Buxrud's *Rare Malts* (Quiller Press, London, 2006) in relation to Diageo's distilleries.

Many monographs about individual distilleries have been published over the years, usually by the distillery's owner. Some are now long out of print; some are more about marketing than history, but some are excellent – I would single out for special praise *Glenfarclas* and *Glengassaugh*, both by Ian Buxton, *The Legend of Laphroaig* by Hans Offringa and Marcel van Gils (Still Publishing, Netherlands, 2007) and *The Island Whisky Trail* by Neil Wilson (NWP, Glasgow, 2003).

In regard to historical overviews, *The Making of Scotch Whisky* by Michael S. Moss (mentioned above) and John Hume (James & James, Edinburgh, 1981) is always on my desk, as is H. Charles Craig's seminal *The Scotch Whisky Industry Record* (Index Publishing, Dumbarton, 1994). I have to admit to consulting my own *Scotch Whisky: A Liquid History* (Cassell, London, 2003) from time to time as well. I owe a special debt to Dr Nicholas Morgan, scholar and sceptic, for many hours of conversation and access to Diageo's extensive archive.

Statistical information comes from the ever helpful Scotch Whisky Association (with special thanks to Campbell Evans, David Williamson and Rosemary Gallagher) and Alan Gray's splendid annual *Scotch Whisky Industry Review*, and also from International Wine & Spirit Research Ltd (IWSR).

For information about production and maturation I am more indebted to individuals than books, and would particularly like to thank Dr Jim Swan, Dr Douglas Murray, Dr Jim Beveridge, Dr Bill Lumsden, Neil Cochrane, Andrew Ford, Paul Lockyer and Philip Murray. The leading current text on the subject is *The Science and Commerce of Whisky* by Ian Buxton and Paul S. Hughes (Royal Society of Chemistry, 2014) – a book I heartily recommend, which stands in the great tradition of scientific texts about Scotch whisky, begun by J.A. Nettleton's *The Manufacture of Whisky and Plain Spirit* (Aberdeen, 1913), which was long the bible for distillers. Nettleton was to production what Barnard was to whisky tourism, which brings us full circle.

As always, friends in the whisky industry have been generous with their time and information, and I am immensely grateful to them. I would particularly like to acknowledge those who checked my text and supplied me with information: Alasdair Anderson (Tamdhu); Russell Anderson (Highland Park); Raymond Armstrong (Bladnoch); Iain Baxter (Inver House); Michael Beamish (Tullibardine); Marcel Bol (Whisky Import Nederland); Johan van Boxmeer (The Old Pipe, St Oedenrode); Carol Brown (Edrington Group); Graham Brown (Tobermory); Steven Burnett

(Macduff); Neil Cameron (Glenturret); Douglas Campbell (Tomatin); John Campbell (Laphroaig); Graham Coull (Glen Moray); David Cox (Macallan); Ronnie Cox (Glenrothes); Jason Craig (Highland Park); Katherine Crisp (Burn Stewart); Francis Cuthbert (Daftmill); Bob Dalgarno (Macallan); David Doig (Fettercairn and Tamnavulin); Gavin Durnin (Loch Lomond); Douglas Fitchett (Glencadam); Callum Fraser (Deanston); Tom Garioch (Tomintoul); Hector Gatt (Glen Scotia); Peter Gordon (Wm Grant & Sons); Donald Graham (Royal Brackla); Micky Heads (Ardbeg); Robbie Hughes (Glengoyne); David Hume (Wm Grant & Sons); Jan Kok (Whisky Import Nederland); Lothar Lange (Whisky-Baritäten); Alastair Longwell (Ardmore); Bill Lumsden (Glenmorangie); Graham Macwilliam (Edrington Group); Des McCagherty (Edradour); Alan McConnochie (BenRiach); Jim McCowan (Cameronbridge); John McLennan (Bunnahabhain); Ian McWilliam (Glenfarclas); Ian Mackie (Whyte & Mackay); Stephen Marshall (John Dewar & Sons); Euan Mitchell (Isle of Arran); Alastair Murphy (North British); Marcin Miller (Glenrothes); Stephanie Mingam (Chivas Brothers); Douglas Murray (Diageo); Karen Murray (Auchentoshan); Stuart Nickerson (Glenglassaugh); Pauline Ogilvy (Glenmorangie); John Peterson (Loch Lomond); Stuart Pirie (Scapa); Robert Ransome (Glenfarclas); Mark Reynier (Bruichladdich); Stuart Robertson (Springbank); Colin Ross (Ben Nevis); Jacqui Seargeant (Bacardi); Graham Sewell (Cameronbridge); Andrew Shand (Speyside); Euan Shand (Huntly); Greg Stables (Glen Grant); David Stewart (Wm Grant & Sons); Andrew Symington (Edradour); Alexander Tweedie (Macallan); Billy Walker (BenRiach); Karen Walker (Inver House); Anthony Wills (Kilchoman); Alan Winchester (Chivas Brothers); Vanessa Wright (Chivas Brothers); Bob Wylie (Morrison Bowmore); special thanks go to Royal Mile Whiskies, especially Arthur Motley for his invaluable help.

Many thanks to Colin Hampden-White for the portrait and Jo Hanley for the bottle photographs. A special thank you to Julie Fergusson, my research assistant. And a huge debt to my editor, Alison Rae.

A Note on the Author

Charles MacLean is a writer whose special subject is Scotch whisky, about which he has published thirteen books to date, including the standard work on whisky brands, *Scotch Whisky* (Mitchell Beazley, 1993) and the leading book on its subject, *Malt Whisky* (Mitchell Beazley, 1997 and 2001; translated into French, German, Swiss, Danish, Italian, Dutch, Polish, Russian and Portuguese), both of which were shortlisted for Glenfiddich Awards. (In addition to whisky, he has published eight further books, mainly on Scottish social history.)

Whisky: A Liquid History was published by Cassell in September 2003 and won 'Wine & Spirits Book of the Year' in the James Beard Awards 2005 and 'Best Drinks Book' in the International Food Media Awards 2005, the most prestigious American gastronomic prize. *MacLean's Whisky Miscellany* (Little Books, 2004, revised 2006 and 2010) followed. In 2007 he was Editor in Chief of the global *Eyewitness Companion to Whisky* (Dorling Kindersley, 2008).

He was founding editor of *Whisky Magazine* and of the Russian magazine *Whisky*, and regularly contributes articles to magazines in Britain, France and China. He is the host of a TV channel (accessed via the internet) dedicated to the world of whisky, www.singlemalt.tv.

He was trained in 'the sensory evaluation of potable spirits' by the Scotch Whisky Research Institute in 1992 and has presented numerous tastings and talks in the U.K. and abroad for whisky companies, corporations, universities and clubs, and on radio and TV.

He is a member of the Judging Panel (Spirits) of the International Wines & Spirits Competition and acts as a consultant to the whisky industry on a variety of matters. Over the past 33 years he has written promotional materials for all the leading companies and brands. In 1992, he was elected a Keeper of the Quaich for 'his services to Scotch over many years' and, in 2009, promoted to 'Master of the Quaich', the industry's highest accolade.

In 2011 he was recruited as script adviser to *The Angels' Share*, a film by Ken Loach and Paul Lafferty, and later played the role of a 'whisky expert' in the film, which won the Jury Prize at the Cannes Film Festival and two Scottish BAFTAs in 2012.

He holds degrees in Art History (St Andrews) and Law (Dundee), is a Writer to Her Majesty's Signet, a Fellow of the Society of Antiquaries of Scotland, a Visiting Lecturer to the Smithsonian Institution, Washington, an Honorary Fellow of Massey College, University of Toronto, and of Champlain College, University of Trent, and a Councillor of the Clan. He lives near Edinburgh with his wife and three sons.

THE SINGLE MALT WHISKY FLAVOUR MAP

The Flavour Map was developed by Diageo's sensory experts in 2007, with help from my esteemed colleague Dave Broom. It presents a new and very helpful way of comparing like with like, and thus offers a guide to styles: if you like malt x, you may like malt y, close by on the map.

Like most great ideas, it is simplicity itself. Malts are plotted on two axes, 'Smoky' to 'Delicate' and 'Light' to 'Rich'.

There is not room to place all malts on the map, so out of deference to its creator, all Diageo's Classic Malts are included, positioned by themselves, while I take responsibility for selecting and locating the others. You may well have your own views about positioning. Good: that's part of the fun of the map!

SMOKY
The single malts in these two quadrants will present discernable levels of peat-derived flavours, from '1960s hospital corridors' to creosote, smoked oysters and smouldering bonfires.

DELICATE
These whiskies are unpeated, even though some might display very faint smoky characteristics. At the lighter end of the scale, you'll find fresh grassy and floral notes; further up the axis the malts become more fruity or malty.

LIGHT
At the left-hand end of the horizontal axis are the fresh-fruity, fragrant, lightly cereal malts. These styles reflect fermentation times and still shape.

RICH
The characteristics found at the right hand end owe more to maturation – whether it be dried fruits, chocolate and cigar boxes (coming from European oak) or deep vanilla and coconut (from American oak).